AF801384

Series in BioEngineering

The Series in Bioengineering serves as an information source for a professional audience in science and technology as well as for advanced students. It covers all applications of the physical sciences and technology to medicine and the life sciences. Its scope ranges from bioengineering, biomedical and clinical engineering to biophysics, biomechanics, biomaterials, and bioinformatics.

Indexed by WTI Frankfurt eG, zbMATH.

Nadia Adrus · Muhammad Aqil bin Mohd Farizal ·
Jamarosliza Jamaluddin ·
Fatria Syaimima bin Syaiful Azim ·
Farahin M. Mizi · Saravanan Nanda Kumar ·
James Jeyadeva Govindasamy

Contact Lenses

Research, Industry, and User Perspectives

Springer

Authors
See next page

ISSN 2196-8861 ISSN 2196-887X (electronic)
Series in BioEngineering
ISBN 978-981-96-3635-8 ISBN 978-981-96-3636-5 (eBook)
https://doi.org/10.1007/978-981-96-3636-5

This work is made possible with the support of the Industry Matching Programme (IMaP) Research Grant Scheme, funded by the Ministry of Higher Education Malaysia, and the Universiti Teknologi Malaysia Fundamental Research Grant (UTMFR Vote: 23H13).

© The Editor(s) (if applicable) and The Author(s), under exclusive license to Springer Nature Singapore Pte Ltd. 2025

This work is subject to copyright. All rights are solely and exclusively licensed by the Publisher, whether the whole or part of the material is concerned, specifically the rights of translation, reprinting, reuse of illustrations, recitation, broadcasting, reproduction on microfilms or in any other physical way, and transmission or information storage and retrieval, electronic adaptation, computer software, or by similar or dissimilar methodology now known or hereafter developed.
The use of general descriptive names, registered names, trademarks, service marks, etc. in this publication does not imply, even in the absence of a specific statement, that such names are exempt from the relevant protective laws and regulations and therefore free for general use.
The publisher, the authors and the editors are safe to assume that the advice and information in this book are believed to be true and accurate at the date of publication. Neither the publisher nor the authors or the editors give a warranty, expressed or implied, with respect to the material contained herein or for any errors or omissions that may have been made. The publisher remains neutral with regard to jurisdictional claims in published maps and institutional affiliations.

This Springer imprint is published by the registered company Springer Nature Singapore Pte Ltd.
The registered company address is: 152 Beach Road, #21-01/04 Gateway East, Singapore 189721, Singapore

If disposing of this product, please recycle the paper.

Nadia Adrus
Faculty of Chemical and Energy Engineering
Universiti Teknologi Malaysia
UTM Johor Bahru, Johor, Malaysia

IJN-UTM Cardiovascular Engineering Centre, Institute of Human Centered Engineering
Universiti Teknologi Malaysia
UTM Johor Bahru, Johor, Malaysia

Jamarosliza Jamaluddin
Faculty of Chemical and Energy Engineering
Universiti Teknologi Malaysia
UTM Johor Bahru, Johor, Malaysia

Institute of Bioproduct Development
Universiti Teknologi Malaysia
UTM Johor Bahru, Johor, Malaysia

Farahin M. Mizi
Faculty of Chemical and Energy Engineering
Universiti Teknologi Malaysia
UTM Johor Bahru, Johor, Malaysia

James Jeyadeva Govindasamy
Alcon (Ciba Vision Johor Sdn Bhd)
Gelang Patah, Johor, Malaysia

Muhammad Aqil bin Mohd Farizal
Faculty of Chemical and Energy Engineering
Universiti Teknologi Malaysia
UTM Johor Bahru, Johor, Malaysia

Fatria Syaimima bin Syaiful Azim
Faculty of Chemical and Energy Engineering
Universiti Teknologi Malaysia
UTM Johor Bahru, Johor, Malaysia

Saravanan Nanda Kumar
Alcon (Ciba Vision Johor Sdn Bhd)
Gelang Patah, Johor, Malaysia

To our parents, spouses, children, siblings, and colleagues

Foreword

As a representative of Alcon, a global leader in eye care, I am honored to contribute to a work that bridges cutting-edge research and practical innovation in the field of contact lenses to the masses. This book represents the fruitful collaboration between Universiti Teknologi Malaysia (UTM) and Alcon Johor also known as CIBA VISION Johor Sdn. Bhd. Together, we have worked to bridge the gap between academic research and industrial application, ensuring that the latest scientific advancements translate into practical solutions that benefit users. With growing demands for innovative, comfortable, and accessible contact lens solutions, this collaboration has emphasized bridging gaps between laboratory discoveries and real-world applications. By addressing both scientific principles and user needs, this book aims to provide a holistic view that benefits researchers, practitioners, and wearers alike. This work is a testament to the power of collaboration and the shared commitment of UTM and Alcon Johor to advancing the field of contact lenses innovations. I am confident that it will serve as a valuable resource for those dedicated to improving vision care and enhancing the lives of users around the world. In Alcon we truly believe, when people see brilliantly, they live brilliantly.

Gelang Patah, Malaysia

James Jeyadeva Govindasamy

Preface

Understanding how contact lenses have evolved from bulky devices to today's cutting-edge innovations is essential in the fast-changing field of ocular technology. This book, *Contact Lenses—Research, Industry, and User Perspectives,* aims to explore this fascinating journey comprehensively, bridging the gap between research advancements, industry practices, and the end-user experience. Specifically, it addresses the interests of material scientists, educators, postgraduate students, manufacturers, and everyday users of contact lenses. The structure of the book has been thoughtfully crafted to accommodate readers from various backgrounds, ensuring accessibility and comprehension for all.

The journey of contact lenses owes much to Otto Wichterle, the father of hydrogel synthesis. I am deeply impressed by Wichterle's passion and how he transformed his idea into an innovative lens machine during the 1960s. Inspired by the smooth parabolic surface formed when sugar mixes into coffee, Wichterle adapted this phenomenon to shape lenses by spinning a liquid polymer. Using handcrafted molds, glass tubing, a toy set, and a bicycle dynamo, he created a lens machine capable of producing 5,500 precise and uniform lenses in just four months. Wichterle's groundbreaking innovation, achieved with such limited resources, is truly extraordinary and remains an enduring source of inspiration for modern advancements.

Building on this remarkable legacy, Chap. 1 explores the historical evolution of contact lenses, highlighting key technological advancements and market dynamics that have shaped the field. Chapter 2 explores the foundational role of hydrogels as the material of choice for contact lenses, focusing on their chemistry, synthesis, and transformative impact on lens performance. Chapter 3 shifts focus to the user, providing a practical guide to navigating options such as therapeutic lenses and specialized designs for astigmatism and presbyopia. Chapter 4 bridges research and manufacturing by addressing critical material properties and performance criteria that define an ideal contact lens, supported by pioneering research and industry standards. For those interested in manufacturing, Chapter 5 examines the industrial landscape, including core processes, quality assurance, and technological advancements. Chapter 6 highlights the essential topic of proper handling and care of contact lenses, addressing common mishandling issues and prevention strategies. Finally,

Chapter 7 looks to the future, exploring emerging technologies like smart lenses, nanotechnology, and expanded healthcare applications, showcasing the potential for transformative impacts in areas such as drug delivery, real-time health monitoring, and augmented reality.

This book represents a significant milestone in my career. With nearly 20 years of experience in UV photopolymerization and functional hydrogels at Universiti Teknologi Malaysia (UTM), I have collaborated with academic colleagues like Dr. Jamarosliza and associate researchers, with a special acknowledgment to Aqil for his dedicated contributions. Farahin and Fatria also contributed significantly to early work on this project, along with our industrial partner Alcon Johor. Key contributors from Alcon Johor include Dr. James, the General Manager, and Saravanan, who is also pursuing his Ph.D. at UTM. Together, we have focused on bridging research and industry to transform hydrogel contact lenses. This effort underscores our commitment to advancing materials and processes that enhance contact lens innovation, providing millions of people with not just vision correction, but also an improved quality of life.

Finally, to our readers—whether academic, professional, or simply passionate about contact lenses—thank you for your curiosity and enthusiasm in exploring this field with us. We hope this book serves as a valuable resource for understanding the current state of contact lenses while inspiring new ideas, research, and innovations in this ever-evolving field. At the same time, we aim to motivate readers to delve deeper into the fascinating world of contact lens advancements. Through this work, we hope to illuminate the path forward in ocular technology, fostering curiosity and encouraging further exploration for contact lenses.

Johor Bahru, Malaysia Nadia Adrus

Acknowledgements The authors would like to express their sincere gratitude to Universiti Teknologi Malaysia Center for Talent and Leadership Development (UTM CTLD) for the invaluable support provided under the Sabbatical Scheme. Special recognition goes to Najwa Wajihah Mohd Rusli, Ariq Ahmad Bowo Leksono, Mohamad Nazri Samin, and Nafiz Waquiddin Mohamad Nazri for their dedication, technical assistance, and unwavering support throughout the research process. Their contributions were instrumental in ensuring the smooth execution of various project components. We are particularly grateful to Van Ho from Seen Eyewear, Johor Premium Outlet (JPO), for their kind assistance in conducting the comprehensive eye tests and fittings. Their expertise and cooperation played a pivotal role in achieving the project's objectives and enhancing the quality of the outcomes.

Competing Interests The authors have no competing interests to declare that are relevant to the content of this manuscript.

Contents

Authors and Contributors

About the Authors

Assoc. Prof. Ir. Dr. Nadia Adrus (Corresponding Author) Faculty of Chemical and Energy Engineering, Universiti Teknologi Malaysia, 81310 Johor Bahru, Johor, Malaysia.

Nadia Adrus is Associate Professor at the Faculty of Chemical and Energy Engineering, Universiti Teknologi Malaysia (UTM). She holds a Ph.D. Degree in Polymer from the University of Duisburg-Essen, Germany. With a specialization in UV photopolymerization and functional hydrogels for diverse applications, she brings almost 20 years of experience in research and teaching to her field. Her role also includes bridging research from the lab to the industry. Since 2016, Dr. Nadia Adrus has closely engaged with Alcon Ciba Vision Johor, to understand their needs, integrating her expertise in developing innovative solutions. She serves in advisory and technical consultancy roles for the photopolymerization of hydrogels used in UV curing of contact lenses. This collaboration underscores her dedication to advancing materials and processes in contact lens technology, facilitating their adoption in real-world applications. Dr. Nadia's academic achievements are distinguished by her numerous publications. Her portfolio includes authored indexed papers, book chapters, and patents, alongside active contributions as Reviewer for papers in reputable journals. Recently certified as Professional Engineer by the Board of Engineer Malaysia (BEM) and Chartered Engineer by the Engineering Council of the United Kingdom (ECUK), Dr. Nadia's dual expertise in academia and industry is widely recognized. e-mail: nadia@utm.my

Muhammad Aqil bin Mohd Farizal Faculty of Chemical and Energy Engineering, Universiti Teknologi Malaysia, 81310 Johor Bahru, Johor, Malaysia.

Aqil Farizal is Research Officer in UTM Technology and Innovation Management Sdn. Bhd., a company under Universiti Teknologi Malaysia (UTM). He graduated from UTM with bachelor's degree in chemical engineering, and previously, he is experienced in conducting research on cellulose dissolution and hydrogel formation during his undergraduate studies. With the knowledge on the fundamentals of polymer, he is currently working on a research project of alternative photoinitiators in contact lens formulation for UV-LED curing system, collaborating with Alcon Ciba Vision in Tanjung Pelepas Port, Johor. During this project, he was involved in analysing contact lens properties and operating the contact lens manufacturing processes in prototype scale. e-mail: m.aqil@graduate.utm.my

Ir. Dr. Jamarosliza Jamaluddin Faculty of Chemical and Energy Engineering, Universiti Teknologi Malaysia, 81310 Johor Bahru, Johor, Malaysia.

Dr. Jamarosliza Jamaluddin is Senior Lecturer in the Department of Chemical Engineering, Faculty of Chemical & Energy Engineering, Universiti Teknologi Malaysia. She is one of the Research Associate in UTM Biopolymer Research Group which focuses on natural fibre, natural rubber, coating and adhesive, hydrogel, and nanocellulose. Dr. Jamarosliza received her bachelor (Chemical-Polymer Engineering) from Universiti Teknologi Malaysia and M.Sc. in Polymer Chemistry from Universiti Putra Malaysia. After completing her master's degree, she pursued her study at Dongguk University, South Korea. Through her passion in material science, particularly in the polymer area, Dr. Jamarosliza's main research area is focused on the polymer adhesive and coating, UV LED technology and photopolymerization, bioplastic and modification of natural fibre. e-mail: jamarosliza@utm.my

Fatria Syaimima bin Syaiful Azim Faculty of Chemical and Energy Engineering, Universiti Teknologi Malaysia, 81310 Johor Bahru, Johor, Malaysia.

Fatria Syaimima Bin Syaiful Azim is Green Enthusiast with more than two years of experience in the research and development field which focused on the use of eco-friendly alternatives and technologies in various applications including contact lens, hydrogel, coating, natural fibre, electronic packaging and bioplastic. He had been actively involved in consultation and collaborative research projects with one of the contact lens manufacturers in Malaysia which prioritized material characteristics and development of contact lens manufacturing. Not only that, he has analytical expertise particularly in handling equipment for characterizing polymer products. He had contributed with several research publications and assisted in research proposals where his team managed to secure a national research grant for a project on the production of bioplastic. Currently, he continues to become Green Devotee as he works as Ooperation Engineer in the oil re-refining company that contributes solutions for a greener environment. e-mail: fatriaazim@gmail.com

Farahin M. Mizi Faculty of Chemical and Energy Engineering, Universiti Teknologi Malaysia, 81310 Johor Bahru, Johor, Malaysia.

Farahin M Mizi graduated with a bachelor's degree in chemical engineering in 2018. Following her interest, she has been working in research and development for almost four years, specializing in polymer, hydrogel as well as in coating and encapsulation. Farahin is highly skilled in formulation development, process optimization and material characterization. She has a demonstrated history of working in consultation with one of Malaysia's contact lens manufacturers, mainly for formulation and process technology development. She has been involved and contributed to several research publications, national conferences, and research collaborations. Presently, she is focusing on the research and development on controlled release technology for environmental applications, specializing in agriculture and public health. e-mail: farahinmizi@utm.my

Saravanan Nanda Kumar Alcon (Ciba Vision Johor Sdn Bhd), No.1, Jalan DPB/ 5, Pelabuhan Tanjung Pelepas, 81560 Gelang Patah, Johor, Malaysia.

Saravanan Nanda Kumar is Permanent Employee and Associate Manager at Alcon Inc., the world's leading eye care company in Johor, Malaysia. This facility produces the renowned monthly Air Optix brand contact lenses. With 17 years of experience, Saravanan specializes in Quality Engineering, Quality Laboratory Testing, Supplier Quality Management, Quality Compliance and Auditing, and Customer Complaints Management. Throughout his career, Saravanan has spearheaded several improvement initiatives, focusing on areas such as supplier quality management, laboratory testing, manufacturing, and quality control process cycle time. His educational background includes a B.Sc. (Hons) in Industrial Chemistry from Universiti Malaysia Sabah and an M.Sc. in Process Integration from Universiti Teknologi Petronas. Currently, Saravanan is pursuing his Ph.D. in Polymer Engineering at Universiti Teknologi Malaysia. His research aims to enhance the quality of contact lenses by optimizing their curing efficiency using UV LED curing. Additionally, he holds a Lean Six Sigma Green Belt certification from Alcon Inc. and is actively working on a project for UV LED curing implementation as part of his ongoing Lean Six Sigma Black Belt certification journey. e-mail: saravanan.nanda_kumar@alcon.com

Dr. James Jeyadeva Govindasamy Alcon (Ciba Vision Johor Sdn Bhd), No.1, Jalan DPB/5, Pelabuhan Tanjung Pelepas, 81560 Gelang Patah, Johor, Malaysia.

Dr. James Govindasamy is currently Site Head for the world's number one eye care company, Alcon Inc.'s contact lens manufacturing plant located in Johor, Malaysia. The plant manufactures exclusively the world-renowned monthly contact lenses, Air Optix product family, for more than 100 countries globally. He has more than 20 years' experience in R&D, manufacturing automation, projects and operations management of budgets exceeding USD100M. Dr. James Govindasamy has a deep interest in industrial automation, robotics and machine learning. He has published various international conference and journal papers in this field as well as an IET book chapter on learning systems for industrial applications. He received his bachelor's degree from Middlesex University, London, for Mechatronics Engineering in 1997 with First Class Honors and a Ph.D. in Controls Engineering from Queen's University of Belfast, in 2004. e-mail: james_jeyadeva.govindasamy@alcon.com

Contributors

Assoc. Prof. Ir. Dr. Nadia Adrus Faculty of Chemical and Energy Engineering, Universiti Teknologi Malaysia, 81310 Johor Bahru, Johor, Malaysia. e-mail: nadia@utm.my

Muhammad Aqil bin Mohd Farizal Faculty of Chemical and Energy Engineering, Universiti Teknologi Malaysia, 81310 Johor Bahru, Johor, Malaysia. e-mail: m.aqil@graduate.utm.my

Ir. Dr. Jamarosliza Jamaluddin Faculty of Chemical and Energy Engineering, Universiti Teknologi Malaysia, 81310 Johor Bahru, Johor, Malaysia. e-mail: jamarosliza@utm.my

Fatria Syaimima bin Syaiful Azim Faculty of Chemical and Energy Engineering, Universiti Teknologi Malaysia, 81310 Johor Bahru, Johor, Malaysia. e-mail: fatriaazim@gmail.com

Farahin M. Mizi Faculty of Chemical and Energy Engineering, Universiti Teknologi Malaysia, 81310 Johor Bahru, Johor, Malaysia. e-mail: farahinmizi@utm.my

Saravanan Nanda Kumar Alcon (Ciba Vision Johor Sdn Bhd), No.1, Jalan DPB/5, Pelabuhan Tanjung Pelepas, 81560 Gelang Patah, Johor, Malaysia. e-mail: saravanan.nanda_kumar@alcon.com

Dr. James Jeyadeva Govindasamy Alcon (Ciba Vision Johor Sdn Bhd), No.1, Jalan DPB/5, Pelabuhan Tanjung Pelepas, 81560 Gelang Patah, Johor, Malaysia. e-mail: james_jeyadeva.govindasamy@alcon.com

Najwa Wajihah Mohd Rusli Faculty of Chemical and Energy Engineering, Universiti Teknologi Malaysia, 81310 Johor Bahru, Johor, Malaysia. e-mail: najwawajihahrusli@gmail.com

Ariq Ahmad Bowo Leksono Faculty of Chemical and Energy Engineering, Universiti Teknologi Malaysia, 81310 Johor Bahru, Johor, Malaysia. e-mail: a.bowo1999@graduate.utm.my

Nafiz Waqiuddin Mohamad Nazri Faculty of Chemical and Energy Engineering, Universiti Teknologi Malaysia, 81310 Johor Bahru, Johor, Malaysia. e-mail: nafiznazri2003@gmail.com

Mohamad Nazri Samin Faculty of Computing, Universiti Teknologi Malaysia, 81310 Johor Bahru, Johor, Malaysia. e-mail: nazrisamin@utm.my

Abbreviations

ADD	Addition
AMA	Allyl Methacrylate
AR	Augmented reality
BC	Base curve
β-CD-crHA	Cross-linked β-cyclodextrin-hyaluronan
BH	Betaxolol hydrochloride
BS	Butyl styrene
BVP	Back vertex power
CAB	Cellulose acetate butyrate
CAD	Computer-aided design
CAGR	Compound annual growth rate
CDVA	Corrected distance visual acuity
CE	Conformité Européenne, or European Conformity
CLARE	Contact lens-induced acute red eye
CoF	Coefficient of friction
CuHCFe	Copper hexacyanoferrate
CVD	Color vision deficiency
CYL	Cylinder
D/N	Dominant
DA	Diacetone acrylamide
DCNVA	Distance-corrected near visual acuity
DED	Dry eye disease
DEGDMA	Diethylene glycol dimethacrylate
DIA	Diameter
DMA	*N,N*-dimethylacrylamide
DOF	Depth of focus
DSAEK	Descemet's stripping automated endothelial keratoplasty
DSS	Dioctyl sulfosuccinate sodium salt
EA	Ethyl acrylate
EDOF	Extended depth of focus
EDS	Ethylene-dimethylsilyl

EGDMA	Ethylene glycol dimethacrylate
EtO	Ethylene oxide
EWC	Equilibrium water content
FDA	Food and Drug Administration
FEA	Finite element analysis
GC	Gas chromatography
GC-MS	Gas chromatography-mass spectrometry
GMA	Glyceryl methacrylate
GNP	Gold nanoparticles
GOx	Glucose oxidase enzyme
GPC	Gel permeation chromatography
H_2O_2	Hydrogen peroxide
HAD	Hyaluronic acid derivative
HCL	Hybrid contact lenses
HDPE	High-density polyethylene
HEMA	2-hydroxyethyl methacrylate
HMI	Human-machine interface
HPMA	Hydroxypropyl methacrylate
I/O	Input/output
IECPS	Integrated eye care professionals
IOP	Intraocular pressure
ISO	International Organization for Standardization
KCS	Keratoconjunctivitis sicca
LC	Liquid crystal
LED	Light-emitting diode
MAA	Methacrylic acid
MFCLs	Multifocal contact lenses
MMA	Methyl methacrylate
MPC	2-methacryloyloxyethyl phosphorylcholine
NCVE	*N*-carboxyl vinyl ester
NF	Norfloxacin
NMR	Nuclear magnetic resonance
NVP	*N*-vinyl pyrrolidone
OCT	Optical coherence tomography
OH	Hydroxyl
PAAm	Polyacrylamide
PC	Phosphorylcholine
PDMS	Polydimethylsiloxane
PEG	Polyethylene glycol
PEGMA	Poly(ethylene glycol) methacrylate
PGP	PDMS–PEGMA
pHEMA	Poly(2-hydroxyethyl methacrylate)
Photo-DSC	Photo-differential scanning calorimetry
PMMA	Poly(methyl methacrylate)
PNIPAAm	Poly(*N*-isopropylacrylamide)

POSTECH	South Korea's Pohang University of Science and Technology
PPy	Polypyrrole
PROSE	Prosthetic replacement of the ocular surface ecosystem
PS	Polystyrene
PSA	Primary spherical aberration
PVA	Polyvinyl alcohol
PVP	Polyvinyl pyrrolidone
PWR	Power/sphere
RF	Radio frequency
RFID	Radio frequency identification
RGP	Rigid gas-permeable
RI	Refractive index
RMS	Root mean squared
SBMA	2-(methacryloyloxy)ethyl] dimethyl-(3-sulfopropyl)
SCL	Soft contact lens
SEM	Standard error of the mean
SiHy	Silicone hydrogel
Si-O	Silicone-oxygen
SSA	Secondary spherical aberration
TCL	Therapeutic contact lenses
TEGDMA	Triethylene glycol dimethacrylate
TiO_2	Titanium dioxide
TPVC	Tris-(trimethylsiloxysilyl) propylvinyl carbamate
TRIS	Tris(trimethylsiloxy)-γ-methacryloxy-propylsilane
USAN	United States adopted names
UTM	Universiti Teknologi Malaysia
UV LED	Ultraviolet light-emitting diode
UVA	Ultraviolet spectrum A (315–400 nm)
UVB	Ultraviolet spectrum B (280–315 nm)
UVC	Ultraviolet spectrum C (100–280 nm)
UV-Hg	Ultraviolet-mercury
VA	Visual acuity
VLCF	Varifocal liquid-crystal Fresnel
VR	Virtual reality
VSOTF	Visual Strehl ratio
WSPI	Water-soluble photoinitiator

List of Figures

List of Tables

Chapter 1
Contact Lenses: A Visionary Journey

1.1 An Insight into Contact Lenses

Contact lenses are thin, clear disks of plastic that are placed directly on the tear film that coats the cornea on the surface of the eye. Contact lenses can be a great alternative to replace the wearing of glasses, that promotes practicality during sports activities and attending outdoor events. In Fig. 1.1, an individual displaying contact lens is portrayed, serving as a representation of the common use of contact lenses for vision correction. On top of this, it is important to acknowledge that contact lenses are considered as medical device, designed to improve the vision of the eyes, either for cosmetic purpose or other suitable reasons.

The lenses are naturally clear in colour, but usually there is a slightest tinge of colour given to the lenses, which enabling the users to see and handle it easily. Currently, contact lenses are either hard or soft, whereby most of the contact lens users prefer the latter for daily usage. This preference for soft lenses is due to their comfort, flexibility, and purpose. Contact lenses are also being worn for purely cosmetic purposes. For instance, coloured contact lenses can be worn to enhance or completely change the hue of the user's eyes. Meanwhile, custom contact lenses help the users to create some special effects on the eyes as can be seen on the big screen. However, all contact lenses either for vision correction or cosmetic purposes require some prescriptions from an eye doctor. This is because common complications such as infection and inflammation could lead to vision loss or blindness to the contact lens users. Besides than vision loss and blindness, other complications of contact lenses will be thoroughly discussed in Chap. 6.

© The Author(s), under exclusive license to Springer Nature Singapore Pte Ltd. 2025

N. Adrus et al., *Contact Lenses*, Series in BioEngineering,
https://doi.org/10.1007/978-981-96-3636-5_1

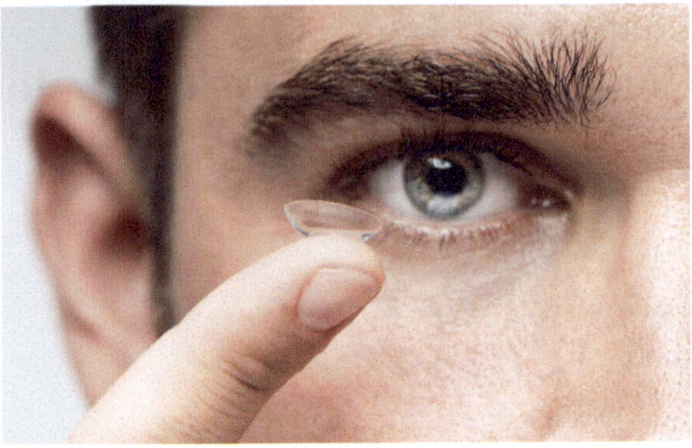

Fig. 1.1 Typical contact lens used for correcting vision (reused from Freepik)

1.2 The Evolutionary Path of Contact Lenses

Throughout history, the evolution of contact lenses has been a fascinating journey, marked by the early theories that laid the groundwork for today's advanced technology. The evolutionary path of contact lens is illustrated in Fig. 1.2, which summarized the historical timeline in the development of contact lenses from the early theories to the current technology of contact lens.

It all starts with the first theory from Leonardo Da Vinci in the year of 1508, which is by immersing the eye in a bowl of water to alter corneal power of the eyes, as illustrated in Fig. 1.3a. His theory did not refer specifically to a mechanism or device for correcting vision, due to his primary interest in learning the mechanisms of accommodation of the eye [1]. In 1636, René Descartes, a French scientist designed a tube that was to be placed directly on the cornea, filled with water with a lens placed on the opposite end, looking like a telescope as depicted in Fig. 1.3b. However, blinking is not possible when using this device, but it proved that the principle of directly neutralizing corneal power is similar to the principle of modern contact lens design [2]. The experiment of this invention continues when Thomas Young, in 1801, came out with a device that is similar to Descartes's tube as shown in Fig. 1.3c, except larger in diameter that can covered the whole eye. Young's invention was more practical because it could be worn securely with a headband and allowed for blinking. However, the device was not meant to correct refractive errors.

With the culmination of early theories, the research on contact lenses became huge in the 1880s. Adolf Gaston Eugene Fick, Eugène Kalt, and August Müller were the scientists that applied the early theories, independently, until they finally created the first glass scleral lenses. As depicted in Fig. 1.4, a scleral lens rests on the sclera, vaulting the whole cornea and limbus and retain a precorneal fluid reservoir. The process of fabricating and fitting contact lenses was first described by Adolf Gaston Eugene Fick in 1888 [3]. In the beginning, Fick described the fitting of afocal scleral contact lens on rabbits, then on himself and finally on a small group of volunteer patients [4]. As for Eugène Kalt, a French ophthalmologist, he found that two patients with keratoconus achieved significant vision improvement by fitting them with afocal

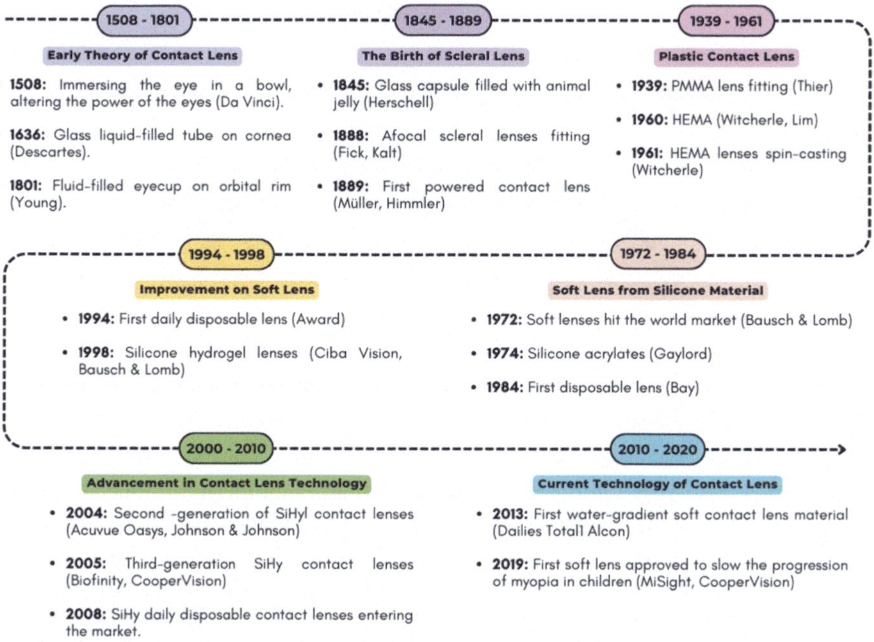

Fig. 1.2 The timeline of contact lens development from early theory to the present technology of contact lens development

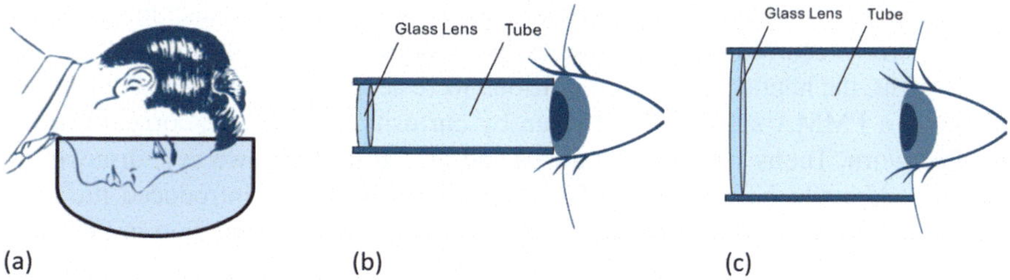

Fig. 1.3 Early theory of correcting vision by **a** Da Vinci; **b** Descartes; and **c** Young

glass scleral lenses [5]. The keratoconus condition is when the cornea gets thinner and bulges outward into a cone shape, causing blurry vision and sensitive to light and glare. Meanwhile, fitting of the first powered contact lenses must be credited to August Müller, which at that time he was a medical student in Kiel University in Germany [6]. In 1889, Müller described in his inaugural dissertation presented to the Faculty of Medicine, where his own high myopia was corrected using a powered scleral contact lens. Interestingly, the lenses worn by Müller were made by Karl Otto Himmler, which was an optical engineer (1841–1903). Therefore, acknowledgement must be given to Himmler as he was the first manufacturer of optically ground contact lenses [7].

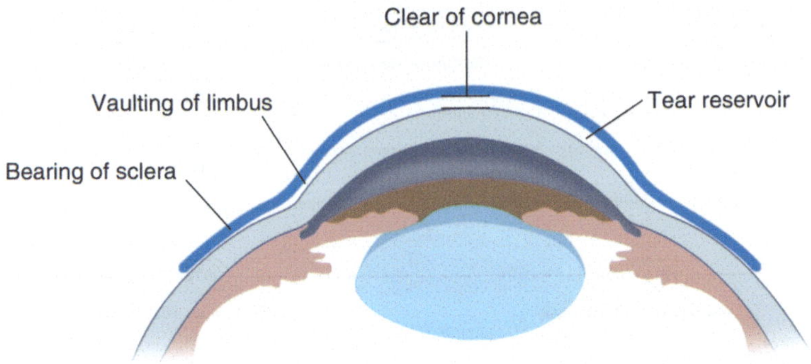

Fig. 1.4 Scleral lens placement on the sclera (reused with permission from Pullum and McMahon [8])

In 1936, a New York Optometrist named William Feinbloom introduced the first plastic contact lens, which retained the scleral lens but featured a clear glass centre [9]. In the same year, the Rohm and Haas company brought polymethyl methacrylate (PMMA) transparent plastic into the USA. Soon after, PMMA were entirely used to fabricate scleral lenses through lathing techniques. The fitting of PMMA lenses were reported earliest in 1939 by Thier and these lenses were claimed to be unbreakable, quicker to manufacture, and lighter about half of the weight of ordinary glass. However, the optical zone of these lenses was required to be polished every 6 months [10]. Later on, in 1948, plastic corneal lenses, also known as rigid lenses, emerged as a new material for contact lenses. The development of corneal contact lenses began when an error occurred in the laboratory by an optical technician, Kevin Tuohy. In this incident, the haptic and corneal portions were accidentally separated during the lathing of a PMMA scleral lens. Driven by curiosity whether the corneal portion could be worn, Tuohy polished the edges and placed it in his own eye, discovering that the lens could be tolerated [11]. In light of this, Tuohy introduced the rigid contact lens by using fully PMMA and filed a patent for his invention in February 1948.

In 1965, contact lenses made from silicone material had begun, marking the introduction of silicone elastomers as the material for 'soft lenses'. Silicone elastomer has high permeability to oxygen and carbon dioxide, minimizing interference with corneal respiration. However, similar to hard lens material, silicone elastomer is inherently dry. Therefore, manufacturing contact lenses from silicone elastomer poses challenges, as its hydrophobic surface must be treated for comfortable wear. Nonetheless, the exact date of commercially available silicon elastomer lenses is unclear, but some patents on this type of lens were active in the mid-1960s to early 1970s [3]. Initially, Otto Wichterle attempted to produce the first world soft lenses fabricated from hydroxythyl methacrylate (HEMA). In the process, he met difficulties in manufacturing the lenses by using cast moulding. Later, he used spin casting technique and eventually managed to produce the lenses, claiming 'the first suitable contact lenses' in late 1961 [12]. Then, Bausch & Lomb acquired the patent to develop soft contact lenses commercially in the USA in 1972. As a result, soft contact lenses manufactured from HEMA became a market success during that time due to the superior comfort and enhanced biocompatibility. In the early days of contact lens development, when HEMA lenses were predominantly thick, the physiological response of the anterior eye while wearing the lens was poor. However, laboratory studies found that this response could be enhanced by making the lenses thinner and increasing their water content.

The drawback of PMMA as the material of contact lens, which is its impermeability to oxygen and carbon dioxide exchanged at the corneal surface as part of aerobic metabolism has been the primary motivation in the development of permeable rigid lens materials. In 1974, Norman Gaylord successfully integrated silicone into the basic PMMA structure, creating a new family of contact lens polymers known as silicone acrylates [13]. Meanwhile in 1980s, disposable lenses were invented as patients would use the same pair of contact lenses until the lenses were damaged or lost, as well as causing discomfort resulting in severe eye reactions. To solve some of these problems, a regular lens replacement was an obvious solution, but the high unit cost of lenses was the major factor of discouragement to the solution of these problems. Michael Bay, an ophthalmologist, who led a group of Danish clinicians and engineers, developed a moulding process that can reduce the cost and can produce multiple individual lens packs [14]. The product is known as 'Danalens' and was recognized to be the first truly disposable lens, which was released into the Scandinavian market in 1984. In the same year, Johnson & Johnson purchased the Danalens technology and completely rebuild the packaging system, lens formulation, and molding technology [14]. As a result, in 1988, an inexpensive weekly-replacement extended-wear lens named Acuvue lens, was released in the USA and worldwide afterwards. Hence, the success of this lens brought Johnson & Johnson to a leading position in the contact lens market at that time.

A Scottish company named Award, developed a technique to reduce the unit cost of a lens, leading to a creation of daily disposable lenses by 1994. This technique was achieved by making the male half of the mould that formed the lens became the lens packaging. Later, in 1996 Bausch & Lomb acquired the Award company, to further advance the technology [3]. In the same year of 1994, Johnson & Johnson launched the 'Premier' daily disposable lens in the UK. They also released the daily disposable lens named '1-Day Acuvue'. Following these development, CIBA Vision entered the market of disposable lens in 1997, with a product called 'Dailies'. After more than a decade of research and development in contact lens, silicone hydrogel lenses with spherical design were introduced in the market in 1998. These lenses, including Focus Night & Day (CIBA Vision) and Purevision (Bausch & Lomb), marked a significant advance in the history of contact lens technology. Due to their superior properties compared to the previous invention, all major contact lens manufacturers swiftly adopted silicone hydrogel (SiHy) lenses that are now available in toric, multifocal designs and various range of replacement modalities.

In the year of 2004, improvements on SiHy contact lenses have been made. These lenses are also known as the second-generation of SiHy contact lenses, where the properties of these lenses are enhanced. The lenses are more comfortable due to their increased water content and reduced moduli, although the first-generation of SiHy contact lenses have higher oxygen permeability [15]. Notably, the reduced moduli of the second-generation SiHy contact lenses prevented several complications such as papillary conjunctivitis and conjunctival epithelial flaps, which caused by relatively high modulus of first-generation silicone hydrogels [16]. Some examples of second-generation SiHy contact lenses are Lotrafilcon B Air Optix™ by Ciba Vision and Galyfilcon A by Johnson & Johnson Vision Care [16]. While for the third-generation of SiHy contact lenses, a unique long-chain siloxane macromer combined with other components was used, resulting lens with high oxygen permeability and a relatively low modulus [16]. The examples of the third-generation of SiHy contact lenses are Comfilcon A Biofinity and Enfilcon A Avaira by CooperVision.

The advancement of contact lens technology continued in 2013 when Alcon introduced their product called 'Dailies Total1' Delefilcon A contact lenses, which are claimed to be the first water-gradient soft contact lens material. The core of the lens is made from a silicone hydrogel material containing 33% water, which can migrate to the outer surface layer, where the water content is 80% [17]. Plus, the low compression modulus on the surface of this lens and the included water gradient technology enables the surface to lubricated and allows for oxygen permeability [18]. Therefore, the symptoms associated with the extended soft contact lens usage such as discomfort due to the friction between the lens surface and the conjunctiva can be prevented. In addition, CooperVision has invented the first soft lens, named MiSight®, that slows the progression of myopia in children. MiSight® 1 day by CooperVision is a daily disposable hydrophilic soft contact lens made from Omafilcon A material, designed to be discarded after each use [19]. It has been clinically proven and FDA-approved for correcting myopia and slowing its progression for children aged 8–12 years with non-diseased eyes. At the start of treatment, these children should have a refraction of − 0.75 to − 4.00 D (spherical equivalent) and astigmatism of not more than 0.75 D. Based on the advancement of silicone hydrogel contact lens, it can be clearly seen that the technology of contact lens is evolving to improve the wearing comfort and treat myopia.

1.3 Forecast and Projected Demand

Ever since the first generation of contact lenses was introduced into the commercial market, their consumption has shown a consistent year-over-year increase. Globally, the market size of contact lenses was estimated at USD 17.14 billion in 2022, and the compound annual growth rate (CAGR) is expected to be 8.9% from 2023 to 2030. The global demand for contact lenses is driven by the increasing cases of refractive errors such as myopia, hyperopia, presbyopia, and astigmatism, as defined in Table 1.1. The increasing demand for contact lenses is driven by the growing ageing population, which is more susceptible to eye disease. According to a recent study by the International Myopia Institute, currently, on average, 30% of the global population is myopic, with projections indicating that nearly 50% will be myopic by 2050 [20]. This rise in myopia is largely influenced by environmental factors, primarily lifestyle changes such as reduced outdoor time and increased near-work activities [21].

As the world is advancing in technology, substantial research and development efforts dedicated to optics and optometry continue to push forward technological advancements. These efforts, including the launch of dynamic soft contact lenses integrated with advanced technology, have significantly propelled the growth of the market. For example, Mojo Vision, a company specializing in micro light-emitting diode (LED) technology for consumer, enterprise, and government applications, made an announcement in 2022 that they have conducted eye testing on their latest prototype—an augmented reality (AR) contact lens. This advanced lens features medical-grade micro-batteries, a micro-LED display, updated hardware, and various technologies including communication, eye tracking, and software. Additionally, the launching of innovative products by the market's leading players is expected to drive the demand for contact lenses. For instance, an eye care product manufacturer named Alcon Vision LLC launched PRECISION1, a daily disposable SiHy lens developed for astigmatic patients in the U.S. The current key players in the contact lens market, originating from the United States of America, Germany, the United Kingdom, France, and Japan, are listed in Table 1.2. These key players are recognized for their robust product portfolio of contact lenses, strategic collaborations, and strong distribution networks.

The increasing popularity of cosmetic and coloured lenses for aesthetic reasons continues to drive the growth of the contact lens market and its global demand. Not only that, more users with active lifestyles like athletes and those involved in physical activities are opting contact lenses due to their convenience and comfort. Nevertheless, the reasons for wearing contact lenses extend beyond these factors. In fact, they can be worn for multiple reasons such as to correct myopia, for aesthetic purposes, and practicality, as outlined in Table 1.3. Furthermore, specific professional occupations rely on contact lenses based on job requirements [22]. According to Fig. 1.5 from 238 subjects, occupations involving near work, computer-related work, and a combination of these categories are prominent among contact lens users. Additionally, roving work, near and distance-related tasks, and near, computer, and roving-related tasks

Table 1.1 Refractive errors that contribute to the increasing global demand for contact lenses

Refractive errors	Definition
Myopia	A vision condition in which near objects can be seen clearly; however, it is difficult to see objects farther away (near-sightedness)
Hyperopia	A vision condition whereby distant objects can be seen more clearly than closer objects
Presbyopia	Difficulty in seeing closer objects for middle-aged and older adults (mid-40s and older)
Astigmatism	A common eye condition arises when the cornea or lens has an irregular shape, causing light to bend unevenly and leading to blurry or distorted vision

Table 1.2 List of the current key players in contact lens market and their origin [20]

Origin	Key players in contact lens market
United States of America	• Abbott Medical Optics, Inc. (Johnson & Johnson) • Alcon Vision LLC • Bausch & Lomb, Incorporated (Valeant International, Inc.) • CooperVision, Inc • Johnson & Johnson Vision Care, Inc. (Johnson & Johnson) • Medennium • STAAR Surgical Company • SynergEyes, Inc • X-Cel Specialty Contacts (Walman Company)
Germany	• ZEISS International (Stiftung)
United Kingdom	• Contamac
France	• Essilor International
Japan	• Hoya Corporation • Seed Co., Ltd

also contribute to the diverse usage of contact lenses across various occupational settings.

In regional perspective, North America dominates the contact lens market with over 38.0% of revenue in 2022, which is predicted to significantly grow in the coming years [20]. Meanwhile the Asia Pacific region is forecasted to achieve a CAGR of 12.0% from 2023 to 2030. According to report by Strait Research, North America's leading revenue share is attributed to the strategic collaborations among market players. For instance, Johnson & Johnson Vision partnered with Menicon, a leading manufacturer of innovative lenses to create therapeutic eyewear in fighting

Table 1.3 Reasons to wear contact lenses

Aspect	Details
Sports	The enhancement to peripheral vision for someone active in outdoor activities to ease activity
Better quality of vision	Improve central and peripheral vision by providing a more natural field of vision, without the obstructions of frame from glasses
Cosmetic	To look better in contacts than glasses and for fashion
Freedom	To prevent discomfort on the nose and ears where the frame of eyeglasses was placed on those regions
Social	To boost self-confidence and eliminate insecurities
Convenience	Contact lenses do not fog up when you go from outside to inside on cold weather
Health	People with certain eye health problems (i.e. keratoconus, anisometropia, aphasia, corneal scarring, trichiasis) can only achieve good vision through contact lens wear
Occupational	Used to correct vision when glasses are unacceptable

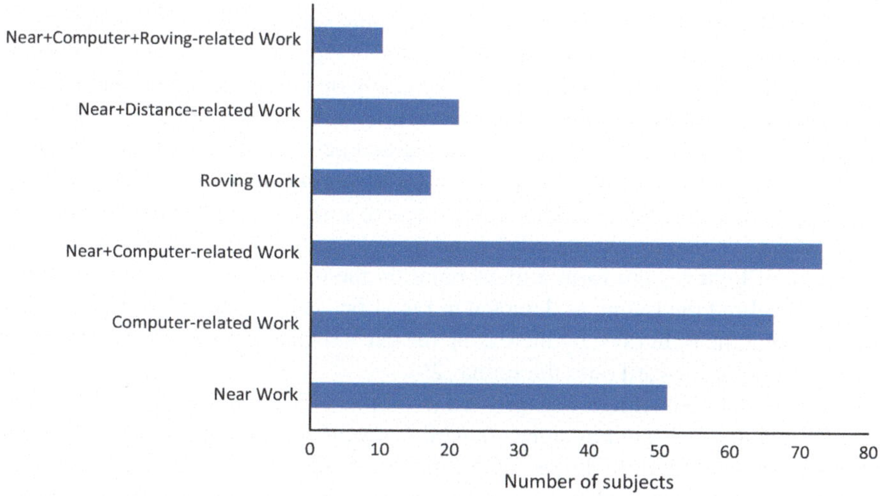

Fig. 1.5 Contact lens usage across occupational categories

myopia among youngsters. In this partnership, Menicon took charge of developing and manufacturing the therapeutic contact lenses, while Johnson & Johnson Vision significantly contributed by broadening the global availability of these lenses, thereby enhancing their market reach and revenue potential [23]. While in Asia Pacific, specifically in Japan, the increasing demand of cosmetic lenses to enhance their eye appearance, is fuelling significant growth in this regional's contact lens market [24]. Besides, the increasing use of contact lenses in China is expected to drive substantial growth in the cosmetic lens market category. Overall, based on the market report from Grandview and Strait Research, the contact lens market is experiencing notable growth, especially in North America and Asia Pacific.

1.4 Improving Vision with Contact Lens

Despite the ongoing evolution of contact lens over time, understanding the underlying science behind their roles in correcting the vision is essential. Contact lenses are considered as specialized medical devices, prescribed for vision correction, for other reasons. They are moulded to conform to the curvature of the cornea of the eye. Today, most of the commercially available contact lenses are soft and flexible, thanks to the use of silicone hydrogel as their primary material. Regardless of their type, contact lenses operate in a similar manner to eyeglasses did to the human eyes, by altering the direction of light rays to focus the light properly onto the retina so that object can appear clearly. Since the lens is placed on the tear fluid of the eye surface, the lens will move naturally with the movement of the eyeball. Hence, this is one of the plus points of contact lens which eyeglasses are unable to provide to the user.

In general, the normal eye problems which require someone to wear contact lenses are either near-sighted or far-sighted, as illustrated in Fig. 1.6. Individuals with normal vision will have the visual image focused precisely on the retina rather than in front of or behind it, as shown in Fig. 1.6a. Meanwhile, in Fig. 1.6b, light rays will form a focus point in front of the retina for near-sighted persons instead of directly on it. Contact lenses are correcting this problem by diverging light rays, which reduces the power focusing of the eye. This moves the eye's focus point backwards onto the retina, where it belongs. On the other hand, people with far-sightedness experience the failure of light rays to form a focus point on the retina, where the visual image is focused behind the retina, as depicted in Fig. 1.6c. Thus, contact lenses solve this by converging the light rays, by increasing the eye's focusing power. This moves the eye's focus point forward onto the retina [25].

The power of contact lens is expressed in diopters (D). For near sightedness, the lens power starts with a minus sign (−), while far sightedness begins with a plus sign (+). When compared to the lenses of eyeglasses, contact lenses have much smaller optic zone about 9 mm, which the central part of the lens contains the corrective power. This is because contact lens rests directly on the cornea, where the optic zone needed is roughly the same as the pupil. However, the placement of lenses for eyeglasses is different, where the lens placement is 12 mm away from the eye's surface, which requires greater optic zone and diameter of lens.

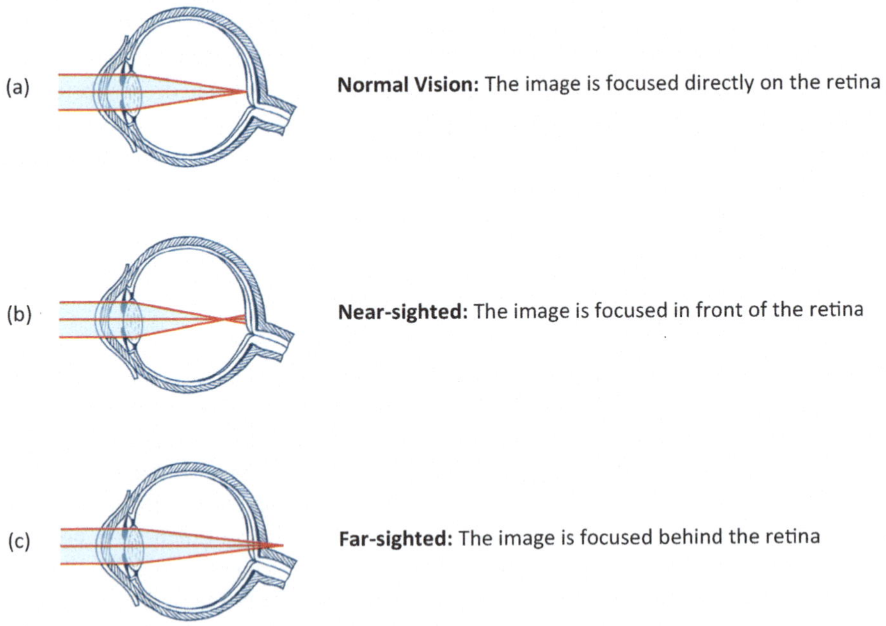

Fig. 1.6 Comparison focused area for **a** normal vision; **b** near-sighted; and **c** far-sighted

Throughout this chapter, the evolution and the significance of contact lenses as a pivotal medical device for vision correction in present days have been explored, highlighting a preference for soft lenses for daily use. The historical journey traced the origin of contact lenses from early conceptualizations by Leonardo Da Vinci to the development of scleral lenses and the adoption of materials like PMMA and HEMA. In later advancements, silicone-based contact lenses emerged as a monumental breakthrough, alongside the introduction of daily disposable lenses. These innovations have further propelled the development of second- and third-generation SiHy lenses, renowned for their superior oxygen permeability that enhances comfort during wear. Additionally, the global market for contact lenses is poised for growth, driven by rising cases of refractive errors and an ageing population, with North America leading the market, followed by the Asia Pacific region. Clearly, the transformative impact of contact lenses in correcting near-sightedness and far-sightedness by effectively directing light rays onto the retina underscores their essential function in vision correction.

In the next chapter, the fundamental role of hydrogel as the basis material for contact lens will be further explained. This includes the hydrogel components, synthesis techniques specific to contact lenses, a comprehensive analysis of their physical, chemical, mechanical, rheological, and biological properties, and the strategic incorporation of silicone to optimize performance in contact lens manufacturing. Understanding these fundamental aspects of hydrogel reveals how advancements in materials science continue to shape the future of contact lens innovation.

References

1. Heitz, R.F., Enoch, J.: Leonardo da Vinci: an assessment on his discourses on image formation in the eye. Adv. Diagnostic Vis. Opt., 19–26 (1987)
2. Enoch, J.M.: Descartes' contact lens. Optom. Vis. Sci. **33**, 77–85 (1956)
3. Efron, N.: Contact Lens Practice-E-Book. Elsevier Health Sciences (2023)
4. Efron, N., Pearson, R.M.: Centenary celebration of Fick's Eine contactbrille. Arch. Ophthalmol. **106**, 1370–1377 (1988). https://doi.org/10.1001/ARCHOPHT.1988.01060140534019
5. Pearson, R.M.: Kalt, Keratoconus, and the contact lens. Optom. Vis. Sci. **66** (1989)
6. Pearson, R.M., Efron, N.: Hundredth anniversary of August Müller's inaugural dissertation on contact lenses. Surv. Ophthalmol. **34**, 133–141 (1989). https://doi.org/10.1016/0039-6257(89)90041-6
7. Pearson, R.M.: Karl Otto Himmler, manufacturer of the first contact lens. Contact Lens Anterior Eye. **30**, 11–16 (2007). https://doi.org/10.1016/J.CLAE.2006.10.003
8. Pullum, K., McMahon, J.: 14 Scleral lenses. Contact Lenses. **289** (2018)
9. Feinbloom, W.: A plastic contact lens*. Optom. Vis. Sci. **14** (1937)
10. Pearson, R.M.: Comments on "Modern scleral contact lenses: a review" [van der Worp et al. (2014)]. Cont. Lens Anterior Eye. **38**, 73–74 (2015). https://doi.org/10.1016/J.CLAE.2014.09.005
11. Braff, S.M.: The Max Schapero lecture: contact lens horizons. Am. J. Optom. Physiol. Opt. **60**, 851–858 (1983). https://doi.org/10.1097/00006324-198310000-00008
12. Wichterle, O.: The beginning of the soft lens. Soft contact lenses Clin. Appl. Technol. (1978)

13. Tighe, B.J.: Extended wear contact lenses. In: Biomaterials and Regenerative Medicine in Ophthalmology, pp. 304–336. Elsevier (2010)
14. Mertz, G.W.: Development of contact lenses. In: Hamano, H., Kaufman, H.E. (eds.) Corneal Physiology and Disposable Contact Lenses, pp. 65–99. Butterworth-Heineman, Boston (1997)
15. Guzin, I., Yunus, K., Ahmet, O., Ceyhun, A., Omer, O., Osman, I.: Comparison of the effects of first and second generation silicone hydrogel contact lens wear on tear film osmolarity. Tear film osmolarity and silicone hydrogel contact lens wear. Int. J. Ophthalmol. **6**, 666–670 (2013)
16. Chou, B., OD, FAAO: The Evolution of Silicone Hydrogel Lenses. https://clspectrum.com/issues/2008/june/the-evolution-of-silicone-hydrogel-lenses. Second generation silicone hydrogels, no surface treatment is applied
17. Tighe, B.J.: A decade of silicone hydrogel development: surface properties, mechanical properties, and ocular compatibility. Eye Contact Lens **39**, 4–12 (2013)
18. Fujimoto, H., Ochi, S., Yamashita, T., Inoue, Y., Kiryu, J.: Role of the water gradient structure in inhibiting thin aqueous layer break in silicone hydrogel-soft contact lens. Transl. Vis. Sci. Technol. **10**, 5 (2021)
19. Ruiz-Pomeda, A., Villa-Collar, C.: Slowing the progression of myopia in children with the MiSight contact lens: a narrative review of the evidence. Ophthalmol. Ther. **9**, 783–795 (2020). https://doi.org/10.1007/s40123-020-00298-y
20. Research, G.V.: Contact Lenses Market Size, Share & Trends Analysis Report 2023–2030
21. Morgan, I.G., Ohno-Matsui, K., Saw, S.-M.: Myopia. Lancet. **379**, 1739–1748 (2012). https://doi.org/10.1016/S0140-6736(12)60272-4
22. Rajan, R., Mahadevan, R., Bhattacharya, P.: Does occupation influence selection of contact lens material? Eye Contact Lens **44**, S43–S47 (2018)
23. Menicon: Menicon announces global collaboration with Johnson & Johnson vision to bring forward therapeutic contact lens related products and services to manage the progression of Myopia in children. https://www.meniconamerica.com/professional/articles/news/menicon-announces-global-collaboration-with-johnson-and-johnson-vision-to-bring-forward-therapeutic-contact-lens-related-products-and-services-to-manage-the-progression-of-myopia-in-children/ (2021)
24. Straits Research: Contact lenses market. Straits Research (2022)
25. Heiting, G.: How do contact lenses work? https://www.allaboutvision.com/contacts/faq/how-contacts-work.htm

Chapter 2
Hydrogel as a Foundational Material for Contact Lens

Hydrogels are a fascinating class of materials meticulously designed for biocompatibility. Because of this, they have emerged as the foundational materials for soft contact lens and continuously revolutionizing in design. Hydrogels are hydrophilic polymers with a strong affinity for water, effectively mimicking the natural hydration of the eye and the cornea's high-water content. Their inherent optical clarity and flexibility make them ideal for addressing the complex demands of ocular applications and vision correction, ensuring both comfort and optimal eye health.

This chapter explores the fundamentals of hydrogels and the unique features that make them ideal for contact lens technology. It emphasizes the importance of tailoring hydrogels to meet specific ocular needs, such as oxygen permeability, biocompatibility and other desirable properties. The chapter also delves into the different types of hydrogels and their synthesis, detailing how monomer formulations are converted to form three-dimensional networks of hydrogel. It further examines how the selection of monomers, crosslinkers, and initiators critically influences the final properties of hydrogels and how these materials are engineered for optimal performance. Finally, the chapter highlights the transformative impact of hydrogels on the evolution of contact lenses, paving the way for innovations in extended wear and next-generation functionalities.

2.1 Understanding Hydrogels: Functionality Tailored for the Ocular Needs

The success of hydrogels in contact lenses stems directly from their unique functionality and chemistry, precisely tailored to meet the specific requirements of the eye. As seen in Fig. 2.1a, a **contact lens** can be viewed as an **extension of the cornea**, where it is applied. As such, it is crucial that the surface and internal properties of

© The Author(s), under exclusive license to Springer Nature Singapore Pte Ltd. 2025
N. Adrus et al., *Contact Lenses*, Series in BioEngineering,
https://doi.org/10.1007/978-981-96-3636-5_2

the hydrogel material used in the contact lens are designed to be compatible with the environment of the corneal epithelium.

Hydrogels are designed to fulfill this compatibility by incorporating key features such **superior optical clarity**, which is essential for providing sharp and unobstructed vision. Their transparent structure allows light to pass through the lens material without significant scattering, ensuring clear vision for the wearer. This clarity is preserved by the stable hydration of the hydrogels, which prevents the formation of micro-dehydration spots that could distort vision. By maintaining consistent hydration levels throughout the day, hydrogels ensure reliable visual performance across various wear schedules, including daily, extended, and continuous use [1]. This combination of clarity and hydration not only enhances comfort but also appeals to users seeking dependable visual acuity.

Since the surface of the eye is highly sensitive and requires continuous hydration and lubrication, hydrogels excel in this aspect due to their high-water content. Unlike most tissues, the cornea lacks its own blood vessels and relies on sufficient oxygenation to support cellular respiration (Fig. 2.1a). Hydrogels thus offer an optimal solution for maintaining both hydration and oxygen permeability, as their **porous, water-rich structure facilitates the passage of dissolved oxygen** through the lens material, ensuring the cornea receives adequate oxygen. Structurally, hydrogels are formed by crosslinking polymer chains, creating a three-dimensional network capable of absorbing and retaining significant amounts of water without dissolving. This network structure enables hydrogels to maintain their water-holding capacity, ensuring they remain intact while allowing them to swell and become soft and flexible.

The presence of **hydrophilic groups** within the polymer matrix attracts and binds water molecules through hydrogen bonding, enabling hydrogels to contain up to 90% water by weight. In the initial designs of soft contact lenses, water content was a key factor in achieving the desired oxygen permeability. The principle was simple—dissolved oxygen could pass through the water molecules within the lens material, allowing more oxygen to reach the cornea. To better understand these relationships, Fig. 2.1b, c, illustrate how copolymers with charged functional groups—such as cationic (NH_3^+) and anionic (COO−) groups—play a crucial role in water absorption by attracting water molecules, ensuring that the lenses maintain their hydrated nature. The high-water content gives them a soft consistency, keeping **lenses moist and reducing friction with the corneal surface**. In doing so, hydrogels effectively address the common issue of dryness associated with traditional lens materials.

To expand on the chemical foundations of these functional groups, Table 2.1 illustrates the **chemical structures** of the **monomers** and **comonomers** commonly used in **hydrogel lens formulations**, highlighting their functional groups and their contributions to **lens characteristics**. Early lenses, such as those made from 2-hydroxyethyl methacrylate (HEMA)-based hydrogels, relied on this principle. The attraction of water molecules to hydroxyl (-OH) groups in HEMA led to water absorption and swelling. Table 2.1 also highlights the roles of various functional groups in enhancing the properties of contact lens materials, including the **carboxyl group** in methacrylic acid (MAA) for ionic properties and water retention, the **lactam group** in N-vinyl

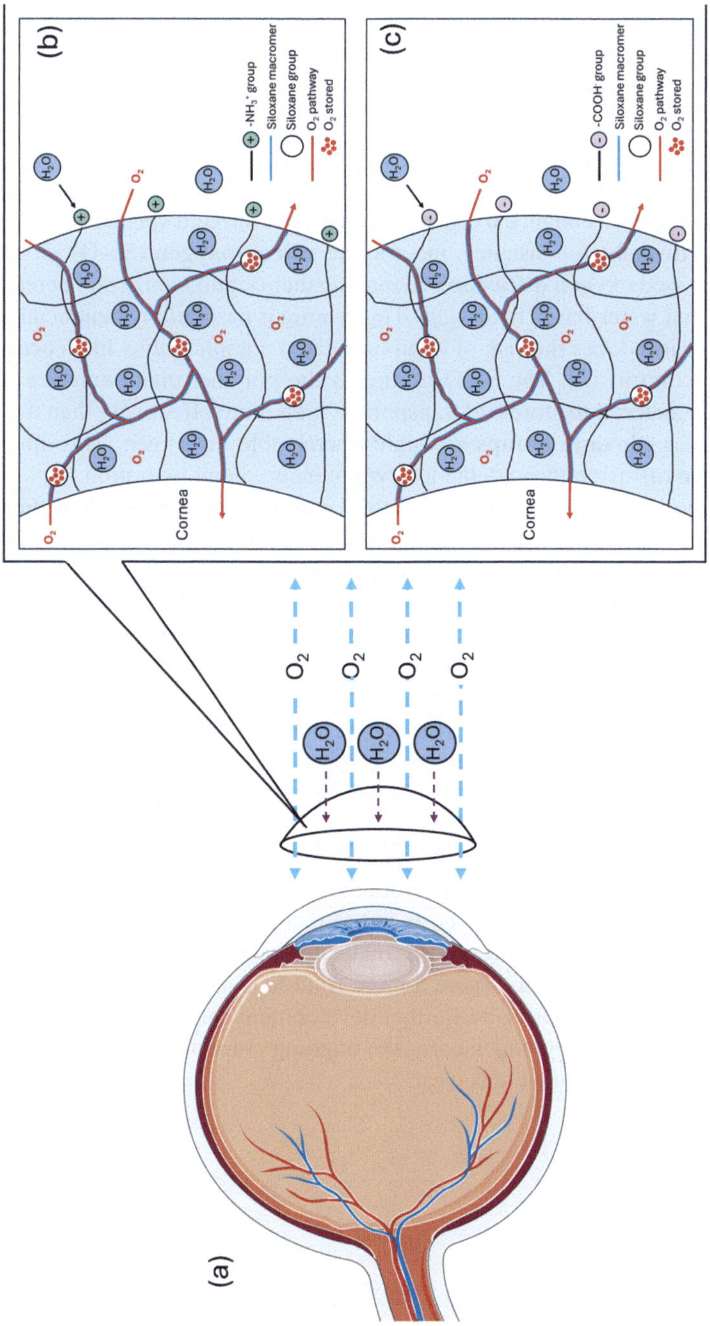

Fig. 2.1 Schematic illustration of hydrogel functionality tailored for ocular needs with **a** contact lens serving as an interface extension to the cornea. **b** Interaction of cationic (**NH_3^+**) and **c** anionic (**COO^-**) functional groups in a hydrogel's porous and water-rich structure facilitates the passage of dissolved oxygen, with silicone macromers enhancing oxygen permeability while maintaining hydration

pyrrolidone (NVP) for increased hydrophilicity and reduced protein buildup, and phosphorylcholine (PC) for improved wettability and moisture retention in extended wear lenses. The use of non-toxic polymers, such as polyethylene glycol (PEG) and polyvinyl alcohol (PVA), further enhances this by minimizing the risk of inflammation or rejection. Together, the high-water content and biocompatible polymers create a smooth, moist-lubricated interface that supports tear film stability and alleviates dryness, making hydrogels ideal for contact lens wear [2, 3].

Modern hydrogel lenses have evolved beyond just increasing water content. They are now designed to enhance **oxygen transmission** through specialized materials. **Silicone hydrogels**, for instance, incorporate **silicon-oxygen (Si–O)** groups, allowing for improved **oxygen diffusion** and making them significantly more breathable than traditional water-based hydrogels. This feature is particularly beneficial for extended wear, as it reduces the risk of hypoxia-related complications like **corneal swelling** and discomfort [4]. The **Si–O** groups in the polymer chain enhance the ability of the lens material to store and transport oxygen more effectively than traditional hydrogels, as **siloxane groups** are highly permeable to oxygen. This allows for greater oxygen diffusion through the lens while maintaining hydration.

The combination of **water retention** from the **cationic** and **anionic functional groups** (Fig. 2.1b, c) and **silicone-based oxygen permeability** ensures that the lens provides a greater supply of oxygen to the cornea, through an additional pathway—not only through the hydrated medium, but also through the incorporation of **silicone groups**. This **copolymer structure** supports **oxygen transmission** through the lens material to reach the corneal surface, improving both comfort and eye health, particularly during extended wear. Additionally, the chemistry behind **silicone hydrogel lenses** makes them more durable and resistant to protein buildup, which helps maintain clarity and comfort by minimizing deposits that cause irritation [5]. This resistance to buildup also improves longevity and hygiene, reducing the need for frequent replacements.

In summary, hydrogels stand out as the preferred material in contact lens technology due to their comfort-enhancing functionality (Fig. 2.1). These materials not only support ocular health and visual acuity but also contribute to the overall satisfaction of contact lens wearers. However, this malleability can also be a disadvantage in certain therapeutic applications where a more rigid lens is needed to maintain a specific shape for corrective purposes. Further development and continuous innovation in contact lens technology underscores the ongoing commitment to improving the safety and comfort of contact lens wear.

Table 2.1 Functional groups and chemical structures of common monomers used in hydrogel lens formulations, with their contributions to lens properties

Monomer	Functional group	Chemical structure	Role in contact lens properties
HEMA	Hydroxyl	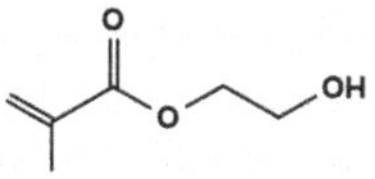	Provides softness and comfort due to moisture content; often used as a base material for soft lenses; limited oxygen permeability
GMA	di-Hydroxyl	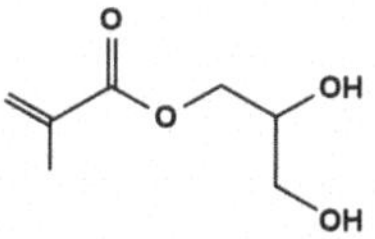	Increase hydrophilicity through the presence of dual hydroxyl groups compared to the single hydroxyl group present in HEMA
MAA	Carboxyl	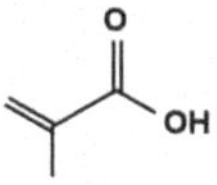	Rarely used alone; enhances water retention and moisture content when combined with HEMA; improves tear film interaction
NVP	Lactam	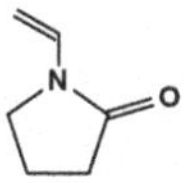	Highly polar, induces hydrophilicity and reduces protein deposition for long-term comfort
DMA	Amine	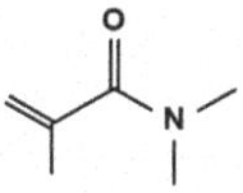	Superior swelling ratio and hydrophilicity compared to NVP
DA	Amine	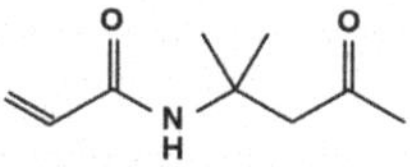	Provides improved oxygen permeability, often used in combination HEMA
PEG	Hydroxyl	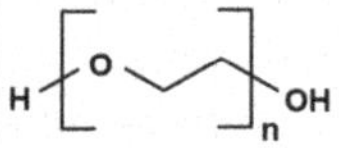	Enhances hydration and tear film stability
PVA	Hydroxyl	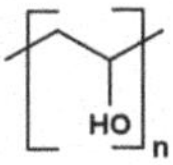	Improves biocompatibility, reducing the risk of inflammation or rejection
MMA	Ester	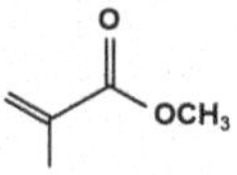	Hydrophobic interaction, enhances lens durability
EGDMA	Ester	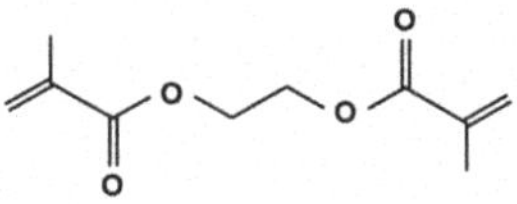	Crosslinking monomer, increases spatial stability and rigidity; forms more compact hydrogel structures

(continued)

Table 2.1 (continued)

Monomer	Functional group	Chemical structure	Role in contact lens properties
MPC	Ester phospholipids, zwitterionic	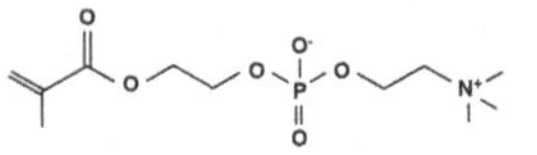	Found in natural membrane lipids, mimics the ocular surface, used together with silicone-based monomer to enhance biocompatibility
TRIS	Siloxane	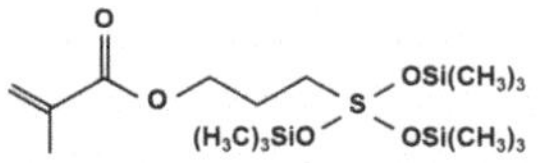	Hydrophobic interaction, significantly enhances oxygen permeability and durability

HEMA—2-Hydroxyethyl methacrylate; GMA—Glyceryl Methacrylate; DMA—(N,N-dimethylacrylamide); DA—diacetone acrylamide; MAA—methacrylic acid; MMA—methyl methacrylate; MPC—2-methacryloyloxyethyl phosphorylcholine; NVP—*N*-vinyl pyrrolidone; PEG—polyethylene glycol; PVA—polyvinyl alcohol; PVP—polyvinyl pyrrolidone; TRIS—tris(trimethylsiloxy)-γ-methacryloxy-propylsilane

2.2 Hydrogel Synthesis and Formulation: From Monomers to Lenses

As introduced in the previous section, the development of fully functional hydrogel contact lenses involves a precise interplay between chemistry and materials science. This section provides an overview of hydrogel synthesis, highlighting the precise chemical reactions and careful formulation guidelines for creating soft contact lenses unveiling the roles of each component in the formulation in determining the final properties of the soft lenses, including its water content, flexibility, and oxygen permeability.

At the core of hydrogel synthesis are **monomers**, which can be visualized as individual building blocks that bond to form long, thread-like polymer chains through a process known as polymerization. In this process, acrylate groups in monomers like **HEMA** or **vinyl groups** in **NVP** react via **free-radical polymerization**, forming covalent bonds that link them into a network. However, simply creating long chains does not result in a hydrogel network. To achieve the desired three-dimensional structure, **crosslinkers** are introduced. These molecules act as bridges, chemically interconnecting the polymer chains at specific points, forming a stable network. Crosslinking can be achieved through chemical methods involving covalent bonds, or physical methods relying on weaker interactions like hydrogen bonding or van der Waals forces. A higher crosslinker concentration generally leads to a tighter, more rigid hydrogel, while a lower concentration results in a softer, more flexible material. **To convert monomers into polymers**, an activation either via heat, light, or chemical reaction is required with the help of an important compound called an **initiator**. When photoinitiators like **Irgacure 1173** are used, they enable the curing of hydrogel contact lenses through photopolymerization using UV light.

The **UV photopolymerization** method, when combined with the cast-molding technique, offers precise control for creating lenses with complex geometries and patterns. While **cast molding** via photopolymerization excels in precision, residues from photoinitiators may impact biocompatibility. On the other hand, **lathe cutting** begins with polymerizing solid rods in bulk, where all components are uniformly mixed. Although bulk polymerization and lathe cutting provide customization options for lenses, they can face scalability challenges. Each synthesis method presents trade-offs in terms of cost, complexity, and scalability, which influences its suitability for contact lens manufacturing. For more detailed information on contact lens manufacturing, please refer to Chap. 5.

Table 2.2 illustrates the HEMA-based monomers and comonomers in contact lens formulations, highlighting how variations in their **chemical functionality** play a key role in determining the **hydrophilicity** and **ionic properties** that categorize lens materials into different Food and Drug Administration (FDA) groups. HEMA has been one of the most widely used monomers in hydrogel contact lens formulations since its discovery by Otto Wichterle in the 1960s and remains a fundamental component [6]. The hydroxyl group in HEMA's chemical structure provides hydrophilicity by enabling hydrogen bonding with water molecules, resulting in soft, flexible, and water-absorbing lenses. On their own, lenses made solely from non-ionic HEMA, such as Polymacon, have relatively low water content.

When HEMA is copolymerized with hydrophilic monomers such as NVP or MAA, the water content increases significantly [7]. NVP contains a lactam ring that significantly enhances hydrophilicity through hydrogen bonding, improving water

Table 2.2 HEMA-based contact lens formulations featuring common monomers and comonomers, and their impact on water content and ionic properties

Chemical composition	Example of contact lens materials	Commercial name	Manufacturer	Water content (%)	FDA group
HEMA	Polymacon	Soflens®	Bausch & Lomb	38	Non-ionic, Low water content
HEMA, NVP	Alphafilcon A	SofLens™ Toric	Bausch & Lomb	66	Non-ionic, High water content
	Hilafilcon A	SofLens™ Comfort	Bausch & Lomb	69	
HEMA, PC	Omafilcon A	Proclear™ Compatible	Cooper Vision	62	
HEMA, MAA	Etafilcon A	Acuvue	Vistakon	58	Ionic, High water content
HEMA, MAA, PVP	Vifilcon A	Focus® Visitint	Ciba Vision	55	
HEMA, DA, MAA	Bufilcon A	Softmate B	Ciba Vision	45	Ionic, Low water content

retention and reducing protein deposition for long-term comfort [8]. This is evident in lenses like Alphafilcon A and Hilafilcon A, which have water contents exceeding 60%. Alternatively, combining HEMA with MAA introduces ionic properties [9]. MAA's carboxylic acid groups ionize in aqueous environments, further increasing water retention and enhancing interaction with the tear film. Lenses such as Etafilcon A and Ocufilcon D, with water contents of 58% and 55%, respectively, benefit from this improved wettability. Another approach involves copolymerizing HEMA with PC, a zwitterionic compound that mimics cell membranes by possessing both positive and negative charges while remaining neutral overall in solution [10]. PC's structure enhances wettability and water retention while maintaining non-ionic characteristics. In formulations like Omafilcon A, the HEMA-PC combination produces high-water-content lenses (62%) that provide superior hydration and comfort [11].

Some formulations combine multiple monomers to optimize water content and resistance to protein deposition. For example, in Vifilcon A (55% water content), HEMA is copolymerized with MAA and polyvinylpyrrolidone (PVP). The combination of ionic MAA and hydrophilic PVP enhances water retention, comfort, and improves replacement frequency [12]. Bufilcon A, on the other hand, is composed of HEMA, MAA, and diacetone acrylamide (DA), and is classified as ionic with mid-water content (45%). DA modifies the hydrophilic-ionic balance, contributing to both comfort and durability, while also reducing the tendency for protein deposition, especially when compared to high-water-content, non-HEMA lenses like Ofilcon A [13].

There are also contact lenses made from non-HEMA formulations, and their materials and key features are summarized in Table 2.3. These lenses use monomers such as NVP, MMA, dimethyl acrylamide (DMA), glyceryl methacrylate (GMA), and PVA, each offering distinct advantages. For instance, lenses like Bausch & Lomb's Medalist and CIBA Vision's Precision utilize NVP and MMA for higher water content and tear resistance [14], while Menicon's Soft Sphere combines NVP and DMA to enhance protein resistance and tear exchange. GMA has two hydroxyl groups on every repeat unit that interact with water molecules, compared to only one in HEMA (Table 2.1), which increases its hydrophilicity. Lenses made with GMA, such as Hioxifilcon A, offer superior clarity but may be prone to dehydration [14], while PVA-based lenses, like Nelfilcon A provide high water content and deposit resistance, ideal for dry-eye patients [15].

The contact lens industry was further revolutionized by the emergence of silicone-based hydrogels, initially developed for rigid gas-permeable (RGP) lenses. In 1974, Gaylord [16] patented a monomer with siloxymethacrylate functionality, which laid the foundation for polymerization of silicon-based monomers through their reactive acrylate groups. This monomer, known as tris(trimethylsiloxy)-γ-methacryloxypropylsilane (TRIS), has since become a proven material for RGP lenses (Fig. 2.2a). The innovation behind silicone-based lenses was driven by the hypothesis that modifying the traditional acrylic hydrogel backbone with siloxane functionality ($-Si(CH_3)_2O-$) could significantly improve oxygen transmission, a key factor in ensuring corneal health during lens wear. Silicone hydrogel lenses typically feature a more complex monomer structure compared to traditional HEMA-based lenses. In

Table 2.3 Non-HEMA-based contact lens formulations highlighting monomer compositions, water content and key features

Chemical composition	Example of contact lens materials	Manufacturer/ Commercial name	Water content (%)	Key features
NVP and MMA	Lidofilcon A	CooperVision's Actifresh 400 UV	70	NVP provides strong attraction to water that give rise to high water content, MAA increases tear resistance and durability
	Vasurfilcon A	CIBA Vision's Precision UV	74	
NVP and DMA	Menicon	Menicon's Soft Sphere and Toric	72	The non-ionic nature and its edge design helps to reduce protein resistant and promote tear exchange
Glyceryl Methacrylate (GMA)	CSI	CIBA Vision's Legacy	58	GMA offers excellent optical clarity and better deposit resistance, though prone to dehydration and fragility
Polyvinyl Alcohol (PVA)	Nelfilcon A	CIBA Vision's Focus Dailies	64	High water content, protein resistant, low lens thickness, good for dry-eye patients

particular, the hydrophobic nature of silicone posed significant challenges, including poor lens wetting and incompatibility with the hydration needs of the ocular environment.

Early attempts to overcome these issues involved direct mixing hydrophilic polymers like HEMA or MAA with silicone to incorporate water-retaining groups, such

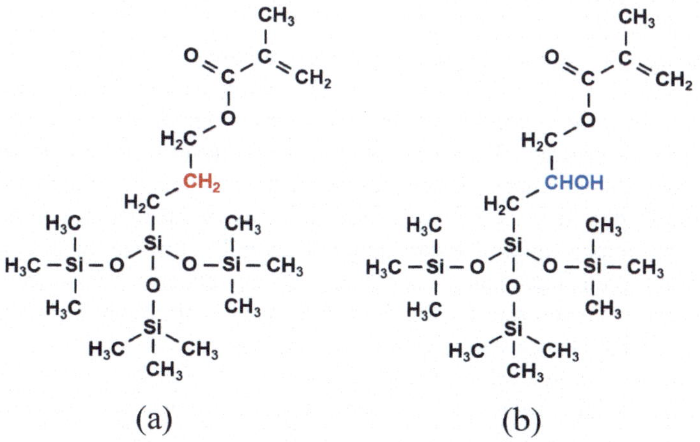

Fig. 2.2 Evolution of TRIS-based monomers: **a** the original structure patented by Gaylord [16], **b** modified structure with a hydroxyl group on the alkyl chain by Tanaka et al. [17]

as hydroxyl (OH). Unfortunately, these efforts often resulted in phase separation, producing lenses that were neither optically clear nor comfortable. This phenomenon arises from limited miscibility between hydrophobic siloxane groups and hydrophilic acrylate groups, driven by unfavorable thermodynamic interactions. Following these challenges, advancements in contact lens formulations focused on modifying the chemistry of organosilicon-containing polymers to enhance compatibility between hydrophilic and hydrophobic components. One such example was the hydrophilic modification of TRIS (Fig. 2.2b), patented by Kyoichi Tanaka at the Toyo Contact Lens Company [17]. His innovative approach involved adding a hydroxyl group to the pendant propyl linkage—a seemingly straightforward chemistry but challenging to implement.

To overcome incompatibility issues, researchers also developed specialized macromonomers, commonly referred to as macromers, to enhance compatibility and functionality in silicone hydrogels [9]. While TRIS consists of silicone rubber segments attached to a modified methyl methacrylate, macromers are essentially large monomers pre-assembled into block copolymers with various functionalities. These macromers are constructed with alternating hydrophilic polyethylene oxide segments and include:

- siloxane and polyether segments ($-OCH_2CH_2-$), with siloxane segments = 25–40, polyether segments = 3–44.
- siloxane and fluorinated ether groups ($-OCF_2CF_2-$), with siloxane segments = 5–100 (especially 14–28), and polyfluoroether segments = 10–30.

In their basic form, each macromer contains two reactive methacrylate groups, one at each chain-end, enabling crosslinking. With a molecular weight typically exceeding 5000, these macromers are designed to impart specific properties to the final polymer, such as improved oxygen permeability and surface wettability. In the case of fluorinated group containing macromer, its low surface energy imparted anti-fouling properties.

Over the decades, numerous macromers have emerged, each with unique chemistry and patent protections. The success of this patent exemplified the potential of macromers in overcoming the challenges posed by the hydrophobic nature of silicone in contact lenses, marking a significant breakthrough in the field of contact lens materials. This macromer-based strategy represents a compelling extension of previous knowledge in overcoming the compatibility of TRIS-like monomers with hydrophilic monomers, as well as enhancing the overall properties of silicone hydrogels. A landmark advancement occurred in 1996 when CIBA Vision (now Alcon) patented a silicone macromer for the development of Lotrafilcon A. In response to market needs, CIBA later established Lotrafilcon B.

As shown in Table 2.4, Lotrafilcon A, Lotrafilcon B and Balafilcon A (Bausch & Lomb) are among the first generation of silicone hydrogel lenses [18]. While Lotrafilcon A and B share the same base formulation utilizing siloxane macromonomers, TRIS, and DMA, Lotrafilcon B features a lower modulus and higher water content than Lotrafilcon A due to differences in composition. In contrast, Balafilcon A incorporates specialized monomers such as tris-(trimethylsiloxysilyl) propylvinyl carbamate (TPVC) and N-carboxyl vinyl ester (NCVE). NCVE introduces ionic functionality, while TPVC provides silicone-based flexibility and high oxygen permeability [18]. These lenses exhibited friction coefficients around 0.4 and a modulus between 1.1 and 1.4 MPa, making them stiffer than HEMA-based materials. They also featured Dk values greater than 100 and relatively low water content—24% for Lotrafilcon A, 33% for Lotrafilcon B, and 36% for Balafilcon A.

The first generation of silicone hydrogel lenses required for plasma treatment to address their low equilibrium water content (EWC) and improve surface wettability. In the second generation, lenses like Senofilcon A (2005, EWC 38%) and Comfilcon A (2006, EWC 48%) were developed with smoother surfaces, eliminating the need for plasma treatments [14, 19]. These lenses exhibited a lower modulus (~0.8 MPa) and significantly reduced friction coefficients (< 0.06), enhancing wettability, comfort, and maintaining high oxygen permeability (100–130 Dk). The third generation, represented by lenses like Enfilcon A [20], retained similar modulus, water content, and oxygen permeability as second-generation lenses, but further refined surface

Table 2.4 Summary of key monomers used in different generation of silicone hydrogel contact lens

Key components	Contact lens materials	Water content/ Oxygen permeability/ Modulus	Notable features
First generation			
DMA, Tris, Siloxane macromer	Lotrafilcon A	24%/140/1.4	High oxygen permeability, rough and hydrophobic surface, stiff materials requiring plasma surface treatment
	Lotrafilcon B	33%/110/1.1	
NCVE, NVP, PBVC, TPVC	Balafilcon A	36%/99/1.1	
Second generation			
HBMA, DMA, VMA TRIS, PBVC, IBMA	Comfilcon A	48%/128/0.75	Improved surface smoothness, lower modulus, enhanced wettability, high oxygen permeability
mPDMS + DMA + HEMA + siloxane macromer + TEGDMA + PVP	Senofilcon A	38%/103/0.72	
Third generation			
NVP, VMA, IBM, TAIC, M3U, FM0411M, HOB	Enfilcon A	46%/100/0.5	Further improved surface smoothness, enhanced comfort, high oxygen permeability, and better wearability
Fourth generation			
Silicone with 2-methacrylphosphoryl choline group	Lehfilcon A	55–100%/154/ 0.6	Gradual increase in water content from core to surface, superior oxygen permeability, Class 1 UV blocking, blue-light filtering

smoothness with friction coefficients around 0.018, leading to improved comfort and ease of wear. The fourth generation, exemplified by Lehfilcon A [21], goes further with exceptional oxygen permeability (154 Dk) and a unique design that gradually increases water content from the core (55%) to the surface (100%). This biomimetic approach enhances hydration while preventing protein deposits, offering additional features like Class 1 UV blocking and blue-light filtering, making Lehfilcon A a leading material for extended and daily wear.

Overall, in this section, we explored the fundamentals of monomers and comonomers used in hydrogel contact lenses, including HEMA-based, non-HEMA, and silicone hydrogels, alongside examples of commercial lenses such as Lotrafilcon A, Balafilcon A, and Senofilcon A. The role of initiators was also briefly discussed, though their influence on the final properties of the lens is not as significant as that of the core monomers and crosslinkers. From this analysis, it becomes clear that material composition, particularly the balance between hydrophilicity, oxygen permeability, and mechanical properties, is crucial to developing comfortable and effective contact lenses. HEMA-based lenses continue to offer proven hydration and flexibility, while non-HEMA formulations and silicone hydrogels have enabled advances in oxygen permeability and long-term wearability, addressing issues like corneal hypoxia.

Looking forward, further innovation in macromer technology and surface treatments could help overcome remaining challenges, such as protein deposition and improving lens comfort for sensitive users. Additionally, evolving techniques like biomimetic designs—such as those seen in the fourth-generation lenses—show promise for better integrating contact lenses with the natural eye environment. In conclusion, the continued refinement of monomer chemistry and polymer formulations, alongside enhancements in manufacturing techniques, will play a pivotal role in the next generation of contact lens development, ensuring even greater performance, comfort, and eye health for wearers.

2.3 Engineering Hydrogel Properties for Enhanced Optical Performance

Did you know that a typical commercially available contact lens is about ten times thicker than the tear film? This significant thickness demands materials with oxygen permeability well beyond what is found in natural tissue. According to mass transfer principles, a thinner lens would allow for improved oxygen flow by minimizing the resistance between surfaces. Yet, designing thinner lenses brings its own set of challenges: manufacturing complexities, handling difficulties due to material softness, and the need to meet exact optical requirements for curvature and power to maintain visual clarity and stability.

Balancing optical clarity, wearer comfort, and tear film stability is therefore a nuanced art in lens design. This section explores two main approaches to achieving this balance. The first approach centers on the engineering aspect of lens itself, where careful selection of specialized monomers and other design strategies are implemented to achieve enhanced lens properties. The second approach involves performance validation through modeling and lens-fitting surveys to assess visual acuity and quantify user experience for long-term wearability. Together, these combined approaches provide insights to validate and refine the engineered properties of contact lenses for enhanced optical performance.

Copolymerization and Its Influence on Critical Properties. Soft contact lenses are primarily made from copolymer-based materials. Researchers are advancing materials that closely align with the physiological needs of the eye through various strategies, including copolymerization with different monomers and crosslinking agents. By integrating hydrophilic or hydrophobic units into a single polymer chain, copolymerization enables precise control over hydrogel durability, water retention, and optical comfort—qualities essential for effective contact lenses.

Research has demonstrated that the oxygen permeability coefficient of hydrogels is directly proportional to their equilibrium water content. However, due to the limited oxygen permeability of water molecules, Dk values in traditional hydrogels typically remain below 40 Barrer, limiting their effectiveness for extended-wear applications. By incorporating hydrophilic co-monomers, manufacturers can improve both water content and permeability, addressing this limitation and supporting tear film stability and ocular comfort for prolonged wear. The following studies compare the effects of simpler silicone-containing monomers, such as TRIS, with more complex silicone macromers like PDMS on soft contact lens properties.

In 2019, Tran's initial study involved the copolymerization of TRIS, a basic hydrophobic monomer, was copolymerized with various amounts of hydrophilic co-monomers, such as DMA, HEMA, and NVP, and crosslinking with ethylene glycol dimethacrylate (EDGMA) [22]. All resulting formulations achieved optical transparency exceeding 95%. With a TRIS at 40 wt% and hydrophilic co-monomers at 60 wt%, the maximum oxygen permeability (74.9 Barrer) and minimum EWC (44.5%) were reached. Adjustments in TRIS and hydrophilic ratios allowed for an oxygen permeability threshold of 70 Barrer with an EWC of 55.9%, enhanced by higher proportions of NVP.

The copolymerization of TRIS with other co-monomers was further investigated by Dhuha and Mutar [23], who prepared three series of formulations: TRIS/AA (Series A), TRIS/AA/DMA (Series B), and TRIS/AA/DMA/NVP (Series C), with hydrophilic co-monomers varying at 70, 80, and 90 wt%. Notably, Series B formulations B1 and B2, with TRIS-to-hydrophilic monomer ratios of 10:90 and 20:80 (excluding NVP), achieved high EWC values above 55% and oxygen permeability over 120 Barrer. Additional formulations by Dhuha et al. replaced DMA with MA in TRIS/MA/HEMA/NVP combinations, yielding the highest oxygen permeability (140 Barrer) and EWC (67%) with a TRIS-to-hydrophilic ratio of 20:80, especially in the absence of MA and with balanced HEMA and NVP ratios [24].

Building on these findings, Tran et al. [25] further refined their approach by substituting TRIS with PGP, a modified polydimethylsiloxane-poly(ethylene glycol) methacrylate (PDMS-PEGMA) macromer, in combination with DMA, NVP, and PEGMA hydrophilic co-monomers in a 35:35:30 ratio. Pure PGP formulations

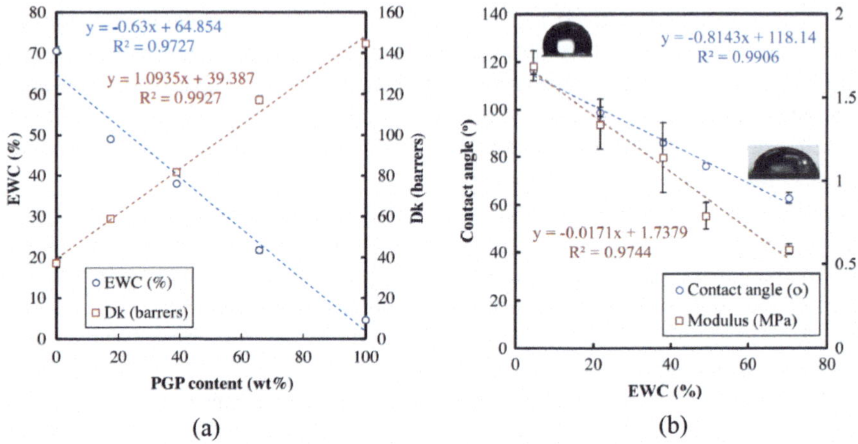

Fig. 2.3 The contact lens properties **a** EWC and oxygen permeability at different concentrations of PGP, and **b** contact angle and modulus at different EWCs (reproduced with permissions from Tran et al. [25])

achieved the highest oxygen permeability ~ 145 Barrer but resulted in a minimal EWC of 4.6%. In contrast, purely hydrophilic formulations yielded an EWC of 70.6% and oxygen permeability of 37.2 Barrer with a modulus of 0.59 MPa. By adjusting macromer/co-polymer ratios, Tran's team achieved a predictive framework, establishing linear relationships with regression fits between 0.97 and 0.99 for critical properties including oxygen permeability, EWC, modulus, contact angle, as shown in Fig. 2.3.

In summary, copolymerization with simpler silicone-containing monomers like TRIS provides a strong foundation for essential properties such as oxygen permeability and optical clarity. Meanwhile, more complex oligomers or pre-polymers like PEGMA or PGP enable greater customization in hydrogel composition, allowing for a fine-tuned balance of water content and permeability to suit varied wearer needs. Findings from Dhuha's [23] study also suggest that formulations with two hydrophilic co-monomers may offer a more favorable balance of properties compared to those with three, as incorporating an additional hydrophilic monomer like NVP may slightly reduce efficiency in achieving optimal oxygen permeability and water retention.

Silicone Monomers and Macromers: Impact of Chain Length and Content. Silicone-based monomers have transformed contact lens technology, providing high oxygen permeability, flexibility, and wearer comfort. Today, more than a dozen silicone hydrogel materials support extended wear and help maintain corneal health, with advancements continually enhancing performance. Increasing the content of silicone macromers, particularly PDMS, within the hydrogel formulation has been shown to directly improve oxygen permeability. For example, Lin et al. [26] demonstrated that raising PDMS content from 80 to 92.3% increased the Dk value from 92 to 110 Barrer, while reducing EWC from 40.8 to 5.5% and modulus from 1.6

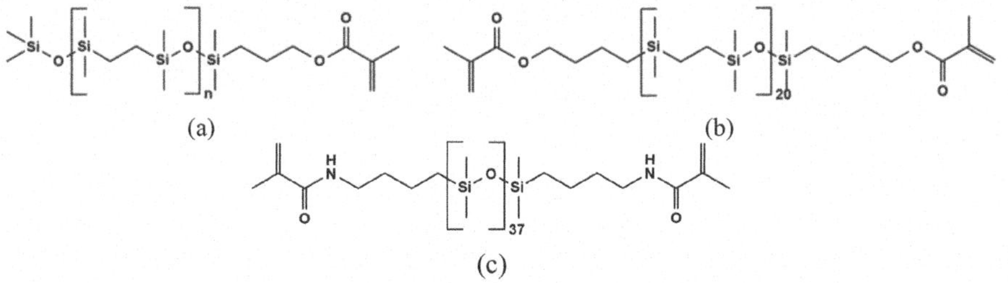

Fig. 2.4 Chemical structure of various silicone-modified macromers **a** mono-ethylenically unsaturated M1-EDS_n, n = 5, 9, 12 and 15 (adapted with permissions from Awasthi et al. [27]), **b** di-ethylenically unsaturated M2-EDS_{20} (adapted with permissions from Awasthi et al. [27]), **c** macromers Ma2D37—patented difunctional macromers of dimethacrylamide of a PDMS diamine [28]

to 0.6 MPa. These results illustrate that higher PDMS content can enhance oxygen permeability while impacting other properties such as elasticity and water retention.

An important study by Awasthi et al. [27] at Bausch & Lomb advanced this work by focusing not only on hydrogel formulation but also on the synthesis of the silicone macromonomers themselves. Awasthi's team synthesized silicone macromers with varying ethylene-dimethylsilyl (EDS). Specifically, a series of M1 mono-ethylenically unsaturated silicone macromers were obtained through anionic ring opening polymerization, yielding a range of EDS substitutions from EDS 6 to EDS 15 (Fig. 2.4a). They also formed M2, di-ethylenically unsaturated silicone macromers with 20 EDS units through cationic polymerization (Fig. 2.4b). Characterization via gel permeation chromatography (GPC), gas chromatography-mass spectrometry (GC–MS), and nuclear magnetic resonance (NMR) confirmed successful synthesis.

Using these synthesized monomers, they created five contact lens formulations, each containing 47.5% hydrophilic monomers (mainly NVP, with DMA and HEMA) and 57.5% hydrophobic components (notably trimethylsiloxysilicate). A photoinitiator (1173) was included at 0.5 wt%. Formulations varied based on the EDS chain length, with entries 1 to 4 incorporating monofunctional silicone macromonomers (EDS 6, 9, 12 and 15) and difunctional macromers (Ma2D37, Fig. 2.5c), while entry 5 used M2EDS20 with an M1 monomer (EDS 6). Results showed that modulus significantly decreased with longer EDS chains, from 1.32 MPa (EDS 6) to 0.82 MPa (EDS 15), and further reduced when M2EDS20 was used. Oxygen permeability (82–89 Barrer) and EWC (41–45%) remained relatively stable across the formulations due to consistent hydrophilic groups and silicone functionality in each entry. This study highlights that while hydrophilic composition and silicone functionality control EWC and Dk, silicone chain length crucially affects stiffness.

Building on these findings, Yokota et al. [29] investigated the effect of chain lengths (1700–7800) in methacryloyl-capped PDMS macromonomers (Fig. 2.5a) within crosslinked copolymers. Their findings emphasized that crosslinker selection and silicone chain length both have a direct impact on gas transmission, with longer PDMS chains generally enhancing oxygen permeability. Then, Tao et al. [30]

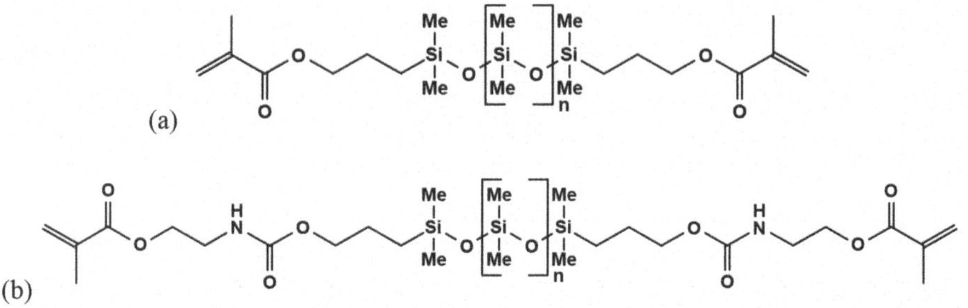

Fig. 2.5 Chemical structure of silicone-modified macromers with two polymerizable methacrylate bonds **a** dimethacryloyl-capped PDMS macromonomers (adapted with permission from Yokota et al. [29]), and **b** dimethacrylamidopropyl-capped PDMS macromers (adapted with permission from Tao et al. [30])

developed high oxygen-permeable silicone hydrogels by copolymerizing PDMS-containing macromonomers with varying siloxane chain lengths (1000–10,000 g/mol, Fig. 2.5b) alongside TRIS and hydrophilic monomers such as NVP, hydroxyl propyl methacrylate (HPMA), and DMA. Oxygen permeability increased up to 243 Barrer at a chain length of 5000 g/mol, but longer chain lengths led to decreased permeability due to phase separation. The EWC exhibited an upward trend, peaking at 25%, while light transmittance declined with longer siloxane chains, with only formulations containing 1000 and 2000 g/mol PDMS maintaining optical clarity above 90% in both dry and swollen states. Tao et al. also noted that longer PDMS chains provided higher mobility and lower stiffness, resulting in decreased modulus, indicating the trade-offs involved in copolymer design for optimal transparency and flexibility.

Research by Wuchte et al. [31] highlights the transformative impact of silicone macromers (Fig. 2.6), particularly methacrylate-functionalized asymmetric silicone macromers, in enhancing contact lens performance. The study focused on reducing phase separation between hydrophilic monomers (DMA) and hydrophobic silicone components (macromer + TRIS). A key discovery was that asymmetric macromers provided stable elastic modulus despite variations in hydrophobic content, unlike symmetric macromers, where increased hydrophobicity compromised modulus, creating comfort trade-offs. Interestingly, asymmetric silicone macromers with fluorinated functionality enhanced optical transmittance to over 85% and improved resistance to surface deposits, though higher fluorine content raised contact angles, slightly reducing wettability. The study also assessed three crosslinkers—EDGMA, and PEGMA with 200 and 550 repeating units—finding that longer crosslinking monomers decreased modulus, producing softer, more flexible lenses for extended wear.

Collectively, with advancements in PDMS content and macromer chain length optimization, these studies reveal significant benefits for enhancing oxygen permeability and flexibility in hydrogels, which are essential for high-performance, comfortable contact lenses suited to extended wear. Awasthi's synthesis of silicone

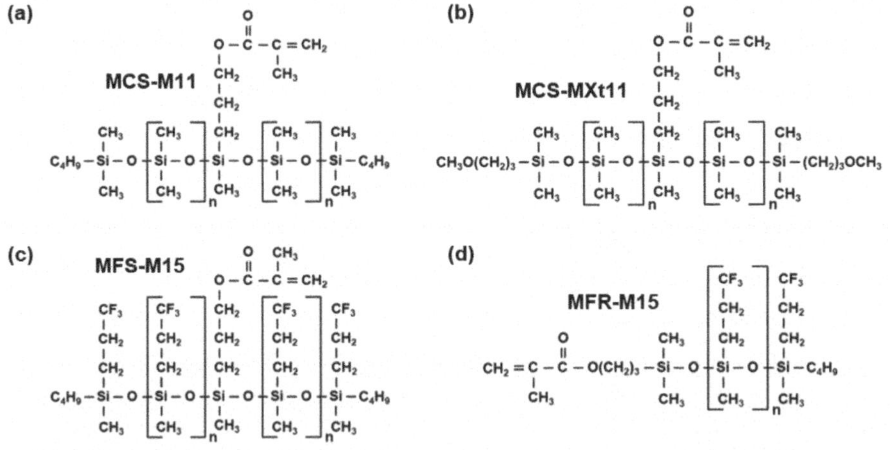

Fig. 2.6 Chemical structure of symmetry functional PDMS macromers with different terminals in (**a**) and (**b**), symmetric PDMS macromer in (**c**) has same terminal as (a) but containing fluorine compounds, and **d** asymmetric PDMS macromer with fluorine (adapted with permissions from Wuchte et al. [31])

monomers with varying chain lengths provides foundational insights, highlighting how silicone chain length specifically impacts stiffness without affecting EWC and Dk. However, a careful balance remains essential, as increased hydrophobicity can trade off comfort and stability, particularly in asymmetric macromers with fluorinated functionality.

Effect of Crosslinking Density on Structural Integrity. Beyond monomer composition, researchers have explored varied crosslinking densities to enhance hydrogel structural integrity and functionality. By adjusting crosslinker content, hydrogels can achieve tailored porosity and structural cohesiveness. Mohammed et al. investigated EGDMA as a crosslinker in TRIS/NVP copolymers, finding that increased EGDMA content reduced free volume, limiting water uptake and thus affecting oxygen permeability [32].

Building on these findings, Wu et al. [33] examined the impact of crosslinking in two series of multicomponent silicone hydrogels using PDMS-1 and PDMS-2, silicone macromonomers with single and double methacrylate terminal functionalities, respectively (Fig. 2.7). In Series One, formulations contained PDMS-1, PDMS-2, or a 50:50 mix, crosslinked with 0.5 phr EGDMA. Series Two used only PDMS-1 with EGDMA at 0.5, 0.75, and 1 phr. Across both series, increased crosslinking density correlated with higher modulus values and reduced oxygen permeability and EWC. PDMS-2, with two polymerizable methacrylate bonds, acted as an intrinsic crosslinker, leading to shorter chain lengths and increased rigidity compared to PDMS-1, which retained a mobile chain end after polymerization. Consequently, Series One exhibited higher crosslink density, resulting in lower oxygen diffusion, reduced water retention, and greater stiffness, emphasizing the role of crosslinker functionality in modifying hydrogel properties.

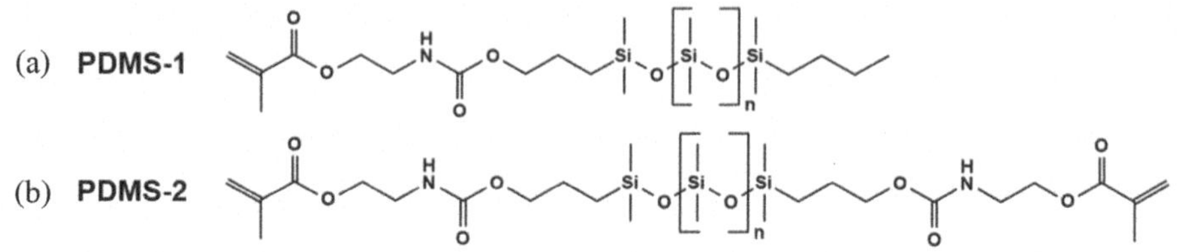

Fig. 2.7 Chemical structures of silicone macromers: **a** PDMS-1, featuring a single methacrylate terminal functionality; and **b** PDMS-2, with double methacrylate terminal functionalities, enabling enhanced crosslinking potential (reused with permissions from Wu et al. [33])

Crosslinking also supports therapeutic applications such as drug delivery, where controlling hydrogel porosity is essential to accommodate different drug sizes and achieve sustained release. Chatterjee et al. reviewed how modifications in crosslinking enhance the hydrogel matrix for controlled drug delivery, establishing contact lenses as a promising platform for targeted ocular treatment [34]. Another crosslinking initiative in ophthalmology, though beyond the scope of this section, involves UV-induced crosslinking of collagen-based material directly applied to the cornea to strengthen tissue structure, especially in cases of keratoconus. Relevant studies include Hovakimyan et al. [35], Sionkowska and Adamiak [36], and Spoerl et al. [37], which explore the potential of UV crosslinking as an emerging treatment approach for enhancing corneal stability.

Hydrogels with lower crosslinker concentrations produce softer, more flexible materials that effectively distribute mechanical stress, reducing corneal irritation for improved comfort. However, reducing crosslinking too much can compromise lens durability and handling. Manufacturers, therefore, adjust crosslinking density to achieve the optimal modulus that balances comfort with structural stability, particularly for specialized lenses like toric lenses for astigmatism, where shape retention is essential. Overall, crosslinking density plays a pivotal role in balancing comfort, permeability, and mechanical strength in hydrogels, allowing lenses to meet both durability and user-specific needs. This controlled approach lays the foundation for versatile, high-performance contact lenses.

Performance Validation Through Modeling and User Experience. While optimizing material properties is essential, understanding how these properties translate into visual performance and comfort is equally critical for successful lens design. Performance validation offers a means to assess these outcomes directly, enabling the refinement of lenses based on real-world user experience.

In their study, Tyagi et al. [38] investigated how contact lens materials affect corneal structure, specifically examining changes in the curvature of the anterior corneal surface and increases in thickness due to wear. Rather than focusing on lens design, this study aimed to understand corneal response to lens materials through a survey of four distinct lens types: three silicone hydrogel lenses and one HEMA-based lens, each in spherical or toric designs with similar center and edge thicknesses (0.11 mm and 0.1 mm, respectively). Findings revealed that the HEMA toric lens with a – 3.00 D prescription caused the most significant corneal thickening, with increases of 1.8% to 5.2% across the central and peripheral cornea after eight hours of wear. In contrast, the same toric design made from silicone hydrogel did not produce similar corneal thickening, highlighting material differences in corneal impact. Additionally, the study observed general anterior corneal flattening across all lenses and a weakly correlated central posterior corneal steepening, suggesting that lens material influences corneal shape changes, with silicone hydrogel lenses demonstrating a reduced impact on corneal thickness compared to HEMA-based lenses.

Complementing this, Marsack et al. [39] conducted a statistical study to identify which optical aberration metrics most accurately predict visual performance, focusing specifically on high-contrast visual acuity. Analyzing 31 different metrics derived from wavefront error data, they found that the visual Strehl ratio (VSOTF) had the strongest correlation with visual acuity, explaining 81% of the observed variation. This finding suggests that VSOTF, a single-value metric, is particularly effective for predicting the impact of wavefront aberrations on visual clarity. Marsack's analysis further revealed that the distribution of aberrations across different Zernike modes influenced visual performance, with modes at the center of each radial order having a more significant effect on acuity than those at the edges. Interestingly, combining certain Zernike modes led to substantial variability in visual acuity even with a constant root mean squared (RMS) error, indicating that total RMS wavefront error alone is insufficient for predicting visual quality. Though the study did not involve clinical trials or direct user experience, its findings provide a generalizable framework for optimizing various optical corrections, including contact lenses, by focusing on wavefront metrics most predictive of visual outcomes.

While Marsack's study [39] used statistical analysis to identify optical aberration metrics that predict visual performance, Plainis et al. [40] took a hands-on approach by exploring how these metrics interact with real-world factors such as pupil size and visual acuity in contact lens wearers. Plainis and colleagues examined the impact of pupil size, spherical aberrations, and binocular vision on visual performance in 12 young participants using center-near multifocal contact lenses (MFCLs). Their study involved Air Optix Aqua Multifocal lenses designed for low, medium, and high near corrections, with experimental conditions adjusted to simulate various visual demands by altering pupil size (3 mm vs. 6 mm) and assessing monocular versus binocular vision. The findings showed that smaller pupil sizes and binocular vision improved both visual acuity (VA) and depth of focus (DOF) at near distances, particularly when using higher-addition lenses. Specifically, VA improved significantly under binocular viewing and with smaller pupils, correlating with enhanced near performance at a 3.0 D defocus. Participants with negative ocular spherical aberrations also benefitted more from the center-near MFCL design, which incorporates negative spherical aberration in its aspheric profile, resulting in superior near VA. This suggests that customizing MFCL designs to account for individual ocular aberrations may optimize both distance and near vision.

Together, these performance validation studies emphasize the need for user-centered metrics to complement material innovations. While engineered properties establish the foundational characteristics of contact lenses, real-world testing—whether through corneal shape analysis, predictive metrics, or simulations of daily visual demands—offers essential insights into how lenses perform over time. This dual approach enables manufacturers to align lens design with user needs, creating lenses that not only meet technical standards but also enhance wearer satisfaction through superior comfort and visual performance.

2.4 The Transformative Impact of Hydrogels on Contact Lens Materials

When pHEMA first emerged, it revolutionized contact lens design, offering a level of comfort previously unseen in rigid lenses. Its ability to retain moisture created a soft, flexible lens that mimicked the eye's natural properties, allowing for extended wear and improved user experience. However, as the demand for superior performance grew, pHEMA revealed its limitations, particularly in oxygen permeability. This spurred decades of innovation, including the introduction of ionic and non-ionic material variations and copolymerization strategies with specialized monomers like DMA, MAA, and NVP, which significantly enhanced lens performance.

Over the years, advancements in silicone hydrogel materials have reshaped the contact lens industry in overcoming the oxygen permeability-water content trade-off. Across four transformative generations, innovations have progressively addressed critical challenges, from early efforts to combat hydrophobicity and stiffness through surface treatments, to the development of inherently wettable materials that eliminated the need for coatings. These advancements have enhanced comfort, accessibility, and lens performance.

Looking ahead, combining the best aspects of these materials to create a revolutionary class of contact lenses progressively **redefined hydrogel performance**. These advancements go beyond merely achieving balances such as hydrophilicity–hydrophobicity, oxygen permeability–wettability, or flexibility–durability to **deliver transformative capabilities**.

Smart lens technologies represent a significant shift in the role of contact lenses, from passive correction to active adaptation. **Photochromic lenses** and **selective UV-blocking lenses** represent two distinct approaches to smart contact lens functionality. Cheng et al. [41] demonstrated one such transformative advancement by developing photochromic contact lenses that dynamically adapt to light conditions. The lenses are composed of a copolymer mixture consisting of **HEMA/MMA/EGDMA** as the crosslinker, and an **acrylate derivative with a photochromic compound as the functional pendant group** (Fig. 2.8). The copolymerization process ensures that the photochromic molecules are covalently bonded to the hydrogel matrix, enhancing durability and optical uniformity. Alternatively, photochromic molecules can be incorporated via physical encapsulation, where they are dispersed and trapped within the porous hydrogel matrix without forming chemical bonds. These lenses exhibit rapid switching between clear and dark states, excellent optical clarity, and high biocompatibility, making them highly adaptable to varying light conditions.

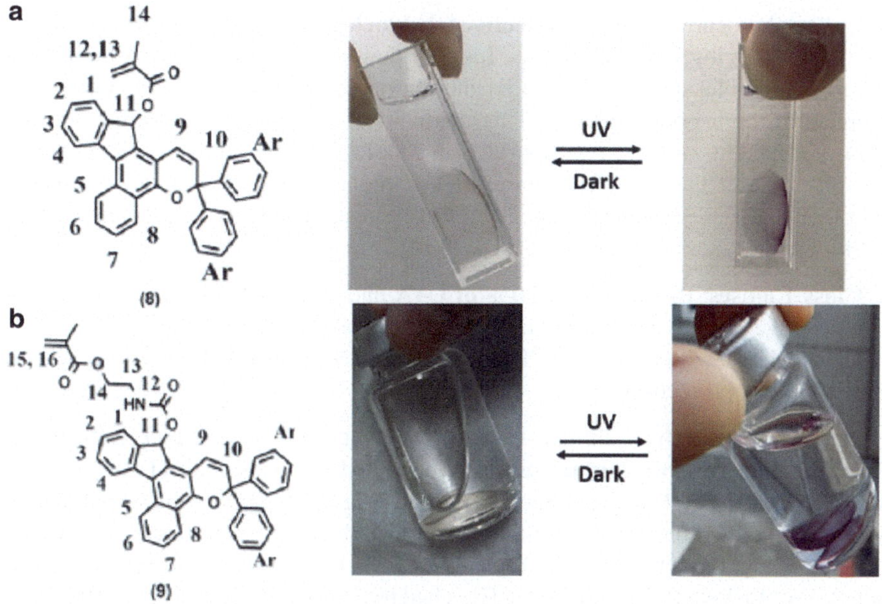

Fig. 2.8 Photochromic transformation of contact lenses copolymer containing: **a** compound 8, and **b** compound 9, both in 0.9 wt% saline solution, showing dynamic color changes under UV light and returning to a colorless state in its absence. Differences in color intensity highlight the distinct photochromic behaviors of the two compounds (adapted with permissions from Cheng et al. [41])

Key findings by Cheng et al. [41] revealed that the performance of these lenses depends on both the choice of photochromic compound and the HEMA content. Lenses with higher HEMA content (e.g., 94 wt%) exhibited faster fading times due to the hydrophilic nature of HEMA, which provides a conducive environment for photochromic transformations. Compound 8 demonstrated faster fading during rapid transitions, while compound 9, with its urethane linkage, benefited from stabilizing hydrogen bonds during slow fading. These photochromic lenses exhibit rapid switching between clear and dark states, slower fading in lens form compared to solution due to structural constraints, excellent optical clarity, and high durability, maintaining functionality over repeated cycles. Such innovations make these lenses highly adaptable to varying light conditions, highlighting the transformative potential of photochromic hydrogels for advanced contact lens applications.

An example of photochromic lens technology being brought to market is **Acuvue Oasys with Transitions Light Intelligent Technology**, a collaboration between Johnson & Johnson and Transitions Optical [42]. Launched in 2019, this was the first photochromic soft contact lens introduced in the U.S. market. These lenses automatically adjust their tint based on light intensity, providing enhanced visual comfort and protection. This commercial success underscores the potential of photochromic hydrogels as practical solutions for real-world applications, further highlighting the transformative impact of such materials on the contact lens industry.

Unlike photochromic lenses, **selective UV-blocking lenses** do not change appearance under varying light conditions. Instead, they provide fixed protection against harmful ultraviolet spectrum B (UVB) and ultraviolet spectrum A (UVA) rays while maintaining high optical clarity. This distinction highlights the different mechanisms and benefits of these two smart lens technologies. UV-blocking contact lenses not only protect the eyes from harmful UV radiation but also shield peripheral light that sunglasses cannot block. The FDA has established two classifications for UV-blocking lenses to ensure standardized performance:

- **Class 1 lenses**: Block at least **90% of UVA** and **99% of UVB** rays, suitable for high UV-exposure environments like beaches and mountains.
- **Class 2 lenses**: Block at least **70% of UVA** and **95% of UVB** rays, designed for general UV protection.

An earlier study by Rahmani et al. [43] evaluated four **commercially available UV-blocking soft contact lenses**: Acuvue Moist, Zeiss CONTACT Day 30 Air Aspheric, Pretty Eyes, and Sauflon 56 UV (Fig. 2.9). Among the tested lenses, Acuvue Moist demonstrated the best overall UV-blocking properties, with 98.78% UVB blocking, though it fell slightly short of FDA Class 1 standards due to insufficient UVA blocking (67.12%). Pretty Eyes had the highest UVB blocking at 99.35%, but its UVA transmittance (34.47%) still fell short of the required threshold. These findings underscore the need for further advancements to enhance UVA protection while maintaining high visible light transmittance for optimal user experience.

The **integration of nearly periodic microsphere arrays within hydrogels** represents a transformative innovation in **achieving Class 1 UV-blocking standards**, effectively blocking UVB (< 1% transmittance) and UVA (< 10% transmittance). Shen et al. [44] pioneered this approach, addressing the limitations of earlier tinted lenses highlighted by Rahmani et al. [43]. These microspheres (190–265 nm),

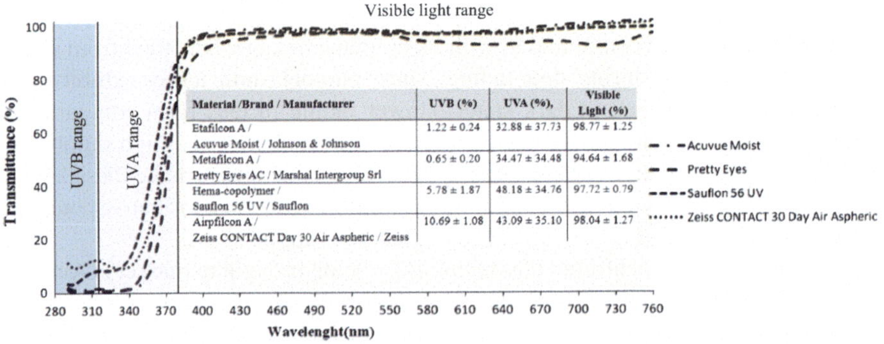

Fig. 2.9 Transmittance curves of four UV-blocking soft contact lenses (Acuvue Moist, Pretty Eyes, Sauflon 56 UV, and Zeiss CONTACT Day 30 Air Aspheric) across UVB (< 315 nm), UVA (315–380 nm) and visible light wavelengths (> 380 nm), with tabulated transmission data for each range (adapted with permissions from Rahmani et al. [43])

consist of a polystyrene (PS)/ PMMA core, an allyl methacrylate (AMA)/ethyl acrylate (EA) interlayer, and a HEMA/EA outer shell. Dispersed within crosslinkable HEMA-DEGDMA (diethylene glycol dimethacrylate) monomer solutions, their high viscosity ensures uniform particle ordering, resulting in high reflectance of short-wavelength light while maintaining transmittance through narrow photonic band gaps. This structural approach enables lenses to scatter and pseudo-absorb harmful UV radiation. Shen's lenses also feature tunable UV protection by adjusting microsphere size and distribution, enhancing adaptability for diverse user needs, with no significant cytotoxicity or impact on cell proliferation [44].

While UV-blocking lenses enhance eye protection, recent innovations have focused on dynamically adapting lens properties to user needs, such as adjustable focal power for presbyopia correction. Wei et al. [45] developed an electronic liquid crystal (LC) contact lens capable of **dynamically switching focal power**. As shown in Fig. 2.10, the relationship between applied voltage and optical power illustrates the lens's ability to smoothly transition between focal states. The optical power varies from approximately − 3.0 D at 0 V to around − 0.5 D at 12 Vrms, showcasing a practical solution for presbyopia correction. The lens operates at low voltages while maintaining precision, making it suitable for dynamic vision adjustment.

In a more recent advancement, Banerjee et al. [46] introduced a **varifocal liquid–crystal Fresnel (VLCF) lens system**, representing a significant leap forward (Fig. 2.11). The VLCF lens stack achieves optical power tunability of ~ 6.5 diopters with ultra-low power consumption (~ 2.6 μW) and a rapid response time (~ 248 ms). Its ultra-thin design (~ 980 μm) and minimized chromatic aberration enhance usability and imaging performance, making it a robust candidate for presbyopia correction. The Fresnel architecture allows for smoother transitions between focal planes and reduces light scattering, overcoming limitations in earlier liquid

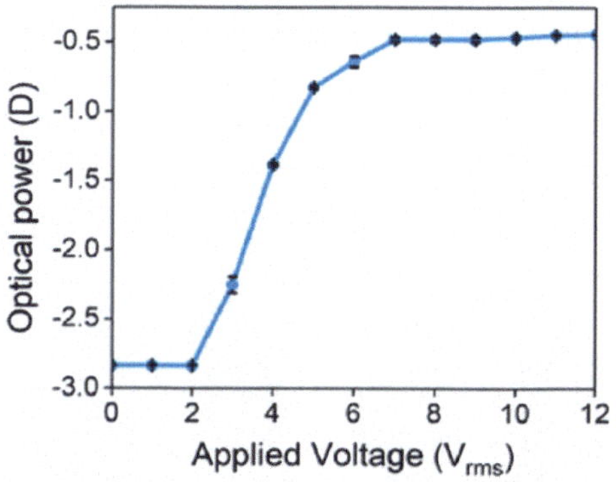

Fig. 2.10 Relationship between applied voltage (Vrms) and optical power (D) in the liquid crystal contact lens (reused with permission from Wei et al. [45]). The lens demonstrates smooth transitions in focal power from − 3.0 D to nearly 0 D as voltage increases, enabling dynamic switching between near and distance vision for presbyopia correction

crystal lenses. These advancements position the VLCF lens system as a transformative innovation in contact lens technology, emphasizing adaptability and energy efficiency.

Hydrogel lenses have **revolutionized therapeutic applications, enabling controlled and sustained ocular drug delivery**. Maulvi et al. [48] demonstrated a significant breakthrough by utilizing polymeric micelles, such as Pluronic® F-68, to encapsulate hydrophobic drugs within contact lenses. This approach achieved prolonged drug release over 72 h without compromising key optical properties, maintaining an impressive 93% light transmittance and manageable swelling essential for visual clarity. Building on this foundation, Wei et al. [49] **advanced drug delivery systems by addressing the critical challenge of oxygen permeability**. Their innovative design embedded a circular-ring layer of Eudragit® S100 in silicone hydrogel lenses, enabling effective delivery of betaxolol hydrochloride (BH) for glaucoma treatment. This design not only ensured sustained therapeutic effects but also maintained adequate oxygen transmission (31.1 ± 3.7 Barrers), making these lenses highly suitable for extended wear without compromising corneal health.

Further advancement by Zhang et al. [47] combined **therapeutic efficacy with intrinsic antifouling properties** through **zwitterionic components**. Specifically, these hydrogels incorporate 2-(methacryloyloxy)ethyl] dimethyl-(3-sulfopropyl) (SBMA) into HEMA and MMA, while also being loaded with norfloxacin (NF). As shown in Fig. 2.11, the NF-loaded hydrogels significantly outperformed their bare counterparts, creating larger bacterial inhibition zones against both *E. coli* (Gram-negative) and *S. aureus* (Gram-positive). Additionally, the lenses retained high optical clarity (94% in the visible range), substantial water retention (> 80%), and mechanical strength (moduli of 0.25–0.42 MPa), ensuring their suitability for extended wear.

Adaptive contact lenses achieve enhanced responsiveness **through the integration of fluorescent probes**, as demonstrated by Yetisen et al. [50]. These lenses act as multifunctional diagnostic devices, enabling real-time monitoring of electrolytes and pH. Silicone hydrogel membranes encapsulate fluorescent probes within sealed microfluidic channels, ensuring tear film stability and high diagnostic precision. The hydrogels' permeability allows detection of electrolytes like Na^+, K^+, and Ca^{2+},

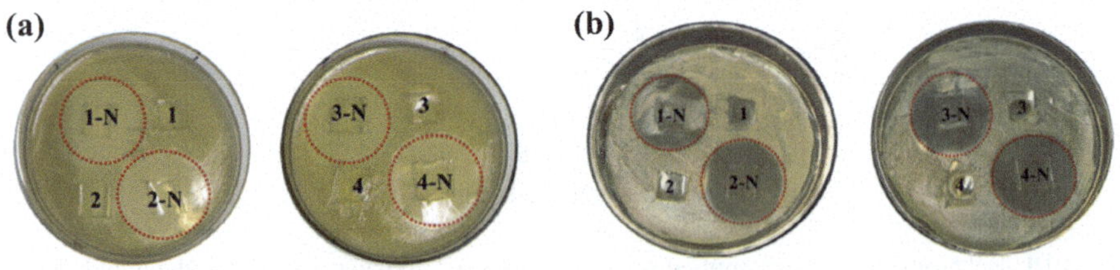

Fig. 2.11 Inhibition zones against **a** *E. coli* and **b** *S. aureus*. Samples 1–4 are bare hydrogels, while 1-N–4-N are NF-loaded hydrogels with descending HEMA and ascending SBMA content. NF-loaded hydrogels show larger inhibition zones, demonstrating enhanced antibacterial activity (reused with permission from Zhang et al. [47])

with sensitivities as low as 0.01 mmol/L for Mg^{2+} and 1 μmol/L for Zn^{2+}. Their biocompatibility ensures prolonged, irritation-free wear, while maintaining optical clarity and mechanical durability for diverse ocular environments. These findings highlight the transformative potential of hydrogels in advancing contact lenses into personalized diagnostic tools.

Wirelessly powered contact lenses to monitor intraocular pressure (IOP) represent another significant advancement in the multifunctionality of hydrogels [51]. Silicone hydrogel lenses serve as the structural base for integrating capacitive sensors and radio frequency identification (RFID)-based energy transfer systems, enabling real-time, non-invasive IOP monitoring. The lenses' mechanical flexibility ensures comfortable long-term wear, while their biocompatibility minimizes risks of irritation. Key innovations include the integration of antenna systems within the hydrogel matrix, which allows efficient wireless power transfer while maintaining safety standards, with negligible temperature increases under RF exposure (< 0.01 °C). These hydrogels pave the way for lenses that bridge therapeutic and diagnostic functionalities, marking a step toward bioelectronic platforms in ocular health.

Sensor integration in hydrogel-based smart lenses has been further investigated by Keum et al. [52] and Elsherif et al. [53], focusing on glucose-monitoring capabilities to enhance the multifunctionality of these platforms. Both studies emphasize the importance of optimizing sensor placement and material properties to balance functionality, optical performance, and user comfort, advancing the role of bioelectronic applications in ocular healthcare. While Elsherif's work reported minor reductions in transparency (from 95 to 92%) due to sensor integration, Keum et al. successfully embedded glucose sensors in hydrogels, achieving 93% transparency, a high equilibrium water content of 33%, and biocompatibility suitable for daily wear. Notably, Keum's smart lens demonstrated the ability to track tear glucose levels in vivo with accuracy comparable to conventional glucometers during key events like insulin administration and anesthesia application [52]. Additionally, the FDA-approved Sensimed Triggerfish® lens, which integrates sensors for intraocular pressure (IOP) monitoring, exemplifies how smart lenses are transitioning from experimental prototypes to real-world applications [54]. When integrating sensors into contact lenses, maintaining flexibility, biocompatibility, and minimal obstruction to vision remains crucial for advancing their practical application [55]. Further details on the Sensimed Triggerfish® lens and its impact can be found in Chap. 7.

Complementing these advancements, Aravind et al. [56] developed **self-moisturizing contact lenses** using hydrogels embedded with capillary microchannels to address the common challenge of dry eye syndrome. The lenses utilize PDMS as a hydrogel base, which is treated with oxygen plasma to achieve a superhydrophilic surface. This treatment enhances capillary-driven fluid flow within 100 μm-wide microchannels, enabling rapid fluid distribution across the lens surface (~ 1.4 s) without external rehydration systems. Importantly, the design preserves optical clarity by concentrating microchannels in peripheral regions, leaving the optical zone smooth and transparent. While challenges remain in optimizing microchannel uniformity due to printing limitations, the approach demonstrates significant potential for creating contact lenses that maintain hydration and enhance user comfort during prolonged wear. These findings highlight the versatility of hydrogels in addressing diverse user needs, from combating dry eye to integrating therapeutic and diagnostic functionalities.

From passive solutions to dynamic and adaptive functions equipped with sensors and diagnostic tools, hydrogel lenses are transforming the boundaries of contact lens technology. No longer confined to vision correction, they now serve as platforms for addressing a wide array of ocular needs. Table 2.5 compiles the role of materials and components in driving innovation across the diverse functionalities discussed in this section.

Through chemical modifications, structural innovations, and integrated electronics, hydrogels have demonstrated their capacity to revolutionize contact lenses. For instance, photochromic pendant groups enable dynamic light adaptation, while microsphere arrays in UV-blocking lenses scatter and pseudoabsorb harmful rays. Zwitterionic additives enhance antifouling properties and, when combined with therapeutic drugs, offer sustained treatment options. Advanced electronic components, such as RFID and capacitive sensors, facilitate self-powered monitoring, and

Table 2.5 Summary of materials, functional roles, and component details driving the transformation of hydrogel contact lenses into dynamics, adaptive and diagnostic platforms

References	Material/ Component	Functionality	Details of component/Techniques
Cheng et al. [41]	Photochromic pendant groups	Light-responsive adaptability	Acrylate derivatives with photochromic pendant groups, e.g., spiropyrans or diarylethenes
Shen et al. [44]	Nearly periodic arrays of Microsphere	UV-blocking through scattering and pseudo-absorption	Well dispersed and uniform ordering of microspheres (190–265 nm) composed of PS/ PMMA core, AMA/EA interlayer, that gives narrow photonic band gaps for UV-blocking
Zhang et al. [47]	Zwitterionic additives	Antibacterial and antifouling properties	SBMA zwitterions integrated with HEMA and MMA
Banerjee et al. [46]	Liquid crystal layers	Adaptive focal power for presbyopia correction	Liquid crystal layers with birefringence and variable optical power tunability
Yetisen et al. [50]	Fluorescent probes	Real-time diagnostics of electrolytes and pH monitoring	Encapsulated probes within silicone hydrogel membranes for stable diagnostic performance
Chiou et al. [51]	Electronic components	Self-powered and wireless IOP monitoring	Capacitive sensors and RFID systems embedded in hydrogels for wireless power and signal transfer
Aravind et al. [56]	–	Self-moisturized lenses through rapid fluid distribution	Microchannels (~ 100 μm wide) formed in PDMS hydrogels through oxygen plasma treatment

microchannel structures created through oxygen plasma treatments tackle hydration challenges without requiring additional material modifications.

2.5 Hydrogels in Focus: A Summary of Their Transformative Role

We have seen how these advancements collectively reshape and redefine the potential of hydrogel materials, transforming them from passive tools of vision correction into dynamic, multifunctional platforms. By seamlessly integrating vision correction, protection, diagnostics, and therapeutic interventions, hydrogels embody the convergence of material science, electronics, and biomedical engineering. Through breakthroughs such as pH-sensitive materials, sustained drug delivery systems, UV-blocking technologies, and sensor-embedded lenses, hydrogels are paving the way for personalized, adaptive, and intelligent ocular solutions.

These innovations highlight hydrogels' unique properties—high-water content, flexibility, and biocompatibility—as foundational for creating lenses tailored to diverse user needs. From zwitterionic materials enabling antifouling and antibacterial applications to photochromic compounds offering dynamic light adaptation, the versatility of hydrogels is evident. Structural advancements like microchannel designs for hydration and cutting-edge technologies, such as glucose-monitoring sensors and wirelessly powered systems for intraocular pressure monitoring, showcase how hydrogels are integrating healthcare and vision care into a single wearable platform.

As we transition into an era of dynamic, biomimetic, and bioelectronic multifunction, hydrogel lenses are evolving into adaptive and intelligent devices capable of responding to environmental and physiological changes. This shift reflects the growing intersection between vision care and health monitoring, positioning hydrogels at the forefront of wearable healthcare technology. The examples discussed in this section represent only a fraction of the rapidly expanding research in this field, with numerous studies and reviews continuing to map the future of hydrogels in contact lenses. By bridging traditional vision correction with personalized healthcare, hydrogels are set to redefine the role of contact lenses in modern medicine, providing transformative solutions that enhance comfort, functionality, and overall ocular health. These advancements underscore the pivotal role hydrogels will play in the next generation of contact lens technologies.

References

1. Sweeney, D.F.: Have silicone hydrogel lenses eliminated hypoxia? Eye Contact Lens **39**, 53–60 (2013). https://doi.org/10.1097/ICL.0b013e31827c7899
2. Li, Q., Cao, Y., Wang, P.: Recent advances in hydrogels for the diagnosis and treatment of dry eye disease. Gels **8**, 1–22 (2022). https://doi.org/10.3390/gels8120816

3. Ferraz, M.P.: Biomaterials for ophthalmic applications. Appl. Sci. **12** (2022). https://doi.org/10.3390/app12125886
4. Holden, B.A., Sweeney, D.F., Vannas, A., Nilsson, K.T., Efron, N.: Effects of long-term extended contact lens wear on the human cornea. Invest. Ophthalmol. Vis. Sci. **26**, 1489–1501 (1985)
5. Ishihara, K., Shi, X., Fukazawa, K., Yamaoka, T., Yao, G., Wu, J.Y.: Biomimetic-engineered silicone hydrogel contact lens materials. ACS Appl. Bio Mater. **6**, 3600–3616 (2023). https://doi.org/10.1021/acsabm.3c00296
6. Wichterle: Hydrophilic gels for biological use. Nat. Publ. Gr. **117**(4706) (1960). https://doi.org/10.1038/185117a0
7. Garrett, Q., Laycock, B., Garrett, R.W.: Hydrogel lens monomer constituents modulate protein sorption. Investig. Ophthalmol. Vis. Sci. **41**, 1687–1695 (2000)
8. Friends, G.D., Künzler, J.F., Ozark, R.M.: High-strength hydrogels based on N-vinyl pyrrolidinone and 4-t-butyl-2-hydroxycyclohexylmethacrylate. J. Biomed. Mater. Res. **26**, 59–67 (1992). https://doi.org/10.1002/jbm.820260106
9. Nicolson, P.C., Vogt, J.: Soft contact lens polymers: an evolution. Biomaterials **22**, 3273–3283 (2001). https://doi.org/10.1016/S0142-9612(01)00165-X
10. Willis S.L., Court J. L., Redman R.P., Wang J.H., Leppard S.W., O'Byrne V.J., Small S.A., Lewis A.L., Jones S.A., Stratford P.W.: A novel phosphorylcholine-coated contact lens for extended wear use. Biomaterials **22**, 3261–3272 (2001). https://doi.org/10.1016/S0142-9612(01)00164-8
11. Riley, C., Chalmers, R.L., Pence, N.: The impact of lens choice in the relief of contact lens related symptoms and ocular surface findings. Contact Lens Anterior Eye. **28**, 13–19 (2005). https://doi.org/10.1016/j.clae.2004.09.002
12. Bleshoy, H., Guillon, M., Shah, D.: Influence of contact lens material surface characteristics on replacement frequency. Int. Contact Lens Clin. **21**, 82–95 (1994). https://doi.org/10.1016/0892-8967(94)90004-3
13. Tomlinson, A., Caroline, P.: Comparative Evaluation of Surface Deposits on High Water Content Hydrogel Contact Lens Polymers. Eye Contact Lens. 16, (1990)
14. González-Méijome, J.M., López-Alemany, A., Almeida, J.B., Parafita, M.A., Refojo, M.F.: Qualitative and quantitative characterization of the in vitro dehydration process of hydrogel contact lenses. J. Biomed. Mater. Res. - Part B Appl. Biomater. 83, 512–526 (2007). https://doi.org/10.1002/jbm.b.30824
15. Bühler, N., Haerri, H.P., Hofmann, M., Irrgang, C., Mühlebach, A., Müller, B., Stockinger, F.: Nelfilcon A, a new material for contact lenses. Chimia (Aarau). **53**, 269–274 (1999). https://doi.org/10.2533/chimia.1999.269
16. Gaylord N.: Oxygen permeable contact lens composition, methods and article of manufacture. https://patents.google.com/patent/US3808178A/en (1974)
17. Tanaka, T.: Phase transitions in gels and a single polymer. Polymer (Guildf). **20**, 1404–1412 (1979). https://doi.org/10.1016/0032-3861(79)90281-7
18. Lira, M., Santos, L., Azeredo, J., Yebra-Pimentel, E., Real Oliveira, M.E.C.D.: The effect of lens wear on refractive index of conventional hydrogel and silicone-hydrogel contact lenses: a comparative study. Contact Lens Anterior Eye. **31**, 89–94 (2008). https://doi.org/10.1016/j.clae.2007.09.001
19. Jones, L., Dumbleton, K.: Soft contact lens fitting. Elsevier Inc. (2019)
20. Luensmann, D., Jones, L.: Protein deposition on contact lenses: The past, the present, and the future. Contact Lens Anterior Eye. **35**, 53–64 (2012). https://doi.org/10.1016/j.clae.2011.12.005
21. Wesley, G., Giedd, B., Hines, B., Bickle, K., Pearson, C., Lorentz, H.: Safety and efficacy of a new water gradient biomimetic monthly replacement spherical contact lens material (Lehfilcon A). Clin. Ophthalmol. **16**, 2873–2884 (2022). https://doi.org/10.2147/OPTH.S362926
22. Tran, N.P.D., Yang, M.C.: Synthesis and characterization of silicone contact lenses based on TRIS-DMA-NVP-HEMA hydrogels. Polymers (Basel). **11** (2019). https://doi.org/10.3390/polym11060944

23. Jawad Hussein, D., Ali Mutar, M.: Synthesis and characterization of new silicone hydrogels contact lenses based on TRIS-DMA-AA-NVP via photo polymerization for biomedical application. Al-Qadisiyah J. Eng. Sci. (2020). https://doi.org/10.30772/qjes.v13i4.727
24. Dhuha, J., Kamel, R.J., Khliwi, F.S., Mutar, M.A.: Synthesis, characterization and optimization of new silicone hydrogel contact lens materials prepared by free radical photopolymerization. J. Mech. Eng. Res. Dev. **44**, 18–32 (2021)
25. Tran, N.P.D., Yang, M.C., Tran-Nguyen, P.L.: Evaluation of silicone hydrogel contact lenses based on poly(dimethylsiloxane) dialkanol and hydrophilic polymers. Coll. Surf. B Biointerfaces. **206**, 111957 (2021). https://doi.org/10.1016/j.colsurfb.2021.111957
26. Lin, C.H., Yeh, Y.H., Lin, W.C., Yang, M.C.: Novel silicone hydrogel based on PDMS and PEGMA for contact lens application. Coll. Surf. B Biointerfaces. **123**, 986–994 (2014). https://doi.org/10.1016/j.colsurfb.2014.10.053
27. Awasthi, A.K., Meng, F.R., Künzler, J.F., Linhardt, J.G., Papagelis, P., Oltean, G., Myers, S.A.: Ethylenically unsaturated polycarbosiloxanes for novel silicone hydrogels: Synthesis, end-group analysis, contact lens formulations, and structure-property correlations. Polym. Adv. Technol. **24**, 557–567 (2013). https://doi.org/10.1002/pat.3115
28. Linhardt Al.: Biomedical Devices United States Patent. https://patents.google.com/patent/US20100317816A1/en (2011)
29. Yokota, M., Ajiro, H., Akashi, M.: Effect of copolymerizing fluorine-bearing monomers on the relationship among internal structure, gas permeability, and transparency in copolymer networks composed of methacrylates and siloxane macromers. J. Appl. Polym. Sci. **127**, 535–543 (2013). https://doi.org/10.1002/app.37813
30. Tao, H., Zhang, X., Sun, Y., Yang, H., Lin, B.: The influence of molecular weight of siloxane macromere on phase separation morphology, oxygen permeability, and mechanical properties in multicomponent silicone hydrogels. Colloid Polym. Sci. **295**, 205–213 (2017). https://doi.org/10.1007/s00396-016-4001-9
31. Wuchte, L., DiPasquale, S., Masterson, A., Vance, A., Goff, J., Arkles, B., Sulaiman, S., Byrne, M.: Characterization and analysis of extended-wear silicone hydrogel contact lenses utilizing novel silicone macromers. J. Biomed. Mater. Res. Part A. **110**, 1512–1523 (2022). https://doi.org/10.1002/jbm.a.37389
32. Mohammed, A.H., Ahmad, M.B., Ibrahim, N.A. et al.: Effect of crosslinking concentration on properties of 3-(trimethoxysilyl) propyl methacrylate/N-vinyl pyrrolidone gels. Chem. Cent. J. **12**, 15 (2018). https://doi.org/10.1186/s13065-018-0379-4
33. Wu, S., Zhang, X., Sun, Y., Yang, H., Lin, B., Han, X., Chen, P.: Study on the influence of crosslinking density and free polysiloxan chain length on oxygen permeability and hydrophilicity of multicomponent silicone hydrogels. Colloid Polym. Sci., 1327–1335 (2021). https://doi.org/10.1007/s00396-021-04850-5
34. Chatterjee, S., Upadhyay, P., Mishra, M., Srividya, M., Akshara, M. R., Kamali, N., Zaidi, Z.S., Iqbal, S.F., Misra, S.K.: Advances in chemistry and composition of soft materials for drug releasing contact lenses. RSC Adv., **10**, 36751–36777 (2020). https://doi.org/10.1039/D0RA06681H
35. Hovakimyan, M., Guthoff, R.F., Stachs, O.: Collagen cross-linking: Current status and future directions. J. Ophthalmol. **2012** (2012). https://doi.org/10.1155/2012/406850
36. Sionkowska, A.: Rheological properties of collagen gels after UV treatment. **415**, 1997 (2020)
37. Spoerl, E., Huhle, M., Seiler, T.: Induction of cross-links in corneal tissue. Exp. Eye Res. **66**, 97–103 (1998). https://doi.org/10.1006/exer.1997.0410
38. Tyagi, G., Collins, M., Read, S., Davis, B.: Regional changes in corneal thickness and shape with soft contact lenses. Optom. Vis. Sci. **87**, 567–575 (2010). https://doi.org/10.1097/OPX.0b013e3181e61b78
39. Marsack, J.D., Thibos, L.N., Applegate, R.A.: Metrics of optical quality derived from wave aberrations predict visual performance. J. Vis. **4**, 322–328 (2004). https://doi.org/10.1167/4.4.8
40. Plainis, S., Ntzilepis, G., Atchison, D.A., Charman, W.N.: Through-focus performance with multifocal contact lenses: effect of binocularity, pupil diameter and inherent ocular aberrations. Ophthalmic Physiol. Opt. **33**, 42–50 (2013). https://doi.org/10.1111/opo.12004

41. Cheng, K.-H., Hsieh, T.-L., Liu, S.-J., Chiang, C.-J., Chen, J.-C.: Synthesis and characterization of indeno-fused naphthopyrans containing methacryloyl and urethane groups for photochromic contact lenses applications. Eur. Polym. J. **211**, 113044 (2024). https://doi.org/10.1016/j.eurpolymj.2024.113044
42. Johnson&Johnson: How does Acuvue® oasys with transitions™ light intelligent technology™ work? https://www.jnj.com/innovation/how-does-acuvue-oasys-with-transitions-light-intelligent-technology-work
43. Rahmani, S., Mohammadi Nia, M., Akbarzadeh Baghban, A., Nazari, M.R., Ghassemi-Broumand, M.: Spectral transmittance of UV-blocking soft contact lenses: a comparative study. Contact Lens Anterior Eye **37**, 451–454 (2014). https://doi.org/10.1016/j.clae.2014.07.011
44. Shen, X., Du, J., Sun, J., Guo, J., Hu, X., Wang, C.: Transparent and UV blocking structural colored hydrogel for contact lenses. ACS Appl. Mater. Interfaces **12**, 39639–39648 (2020). https://doi.org/10.1021/acsami.0c10763
45. Wei, Z., He, M., Li, S., Zhang, Z., Yang, L., Chen, Y., Qu, S., Wang, Q., Qian, J.: Electrically tunable liquid crystal lens with varifocal capability. ACS Appl. Opt. Mater. (2024). https://doi.org/10.1021/ACSAOM.4C00339/ASSET/IMAGES/MEDIUM/OT4C00339_0010.GIF
46. Banerjee, A., Ghosh, C., Karkhanis, M.U., Deshpande, A., Pourshaban, E., Majumder, A., Kim, H., Mastrangelo, C.H.: Refractive-type varifocal liquid-crystal Fresnel lenses for smart contacts. Opt. Express **31**, 17027 (2023). https://doi.org/10.1364/oe.489093
47. Zhang, J., Qian, S., Chen, L., Wu, M., Cai, Y., Mou, X., Feng, J.: Antifouling and antibacterial Zwitterionic hydrogels as soft contact lens against ocular bacterial infections. Eur. Polym. J. **167**, 111037 (2022). https://doi.org/10.1016/j.eurpolymj.2022.111037
48. Maulvi, F.A., Parmar, R.J., Desai, A.R., Desai, D.M., Shukla, M.R., Ranch, K.M., Shah, S.A., Shah, D.O.: Tailored gatifloxacin Pluronic® F-68-loaded contact lens: Addressing the issue of transmittance and swelling. Int. J. Pharm. **581** (2020). https://doi.org/10.1016/j.ijpharm.2020.119279
49. Wei, Y., Hu, Y., Shen, X., Zhang, X., Guan, J., Mao, S.: Design of circular-ring film embedded contact lens for improved compatibility and sustained ocular drug delivery. Eur. J. Pharm. Biopharm. **157**, 28–37 (2020). https://doi.org/10.1016/j.ejpb.2020.09.010
50. Yetisen, A.K., Jiang, N., Castaneda Gonzalez, C.M., Erenoglu, Z.I., Dong, J., Dong, X., Stößer, S., Brischwein, M., Butt, H., Cordeiro, M.F., Jakobi, M., Hayden, O., Koch, A.W.: Scleral lens sensor for ocular electrolyte analysis. Adv. Mater. **32** (2020). https://doi.org/10.1002/adma.201906762
51. Chiou, J.C., Hsu, S.H., Liao, Y.T., Huang, Y.C., Yeh, G.T., Kuei, C.K., Dai, K.S.: Toward a wirelessly powered on-lens intraocular pressure monitoring system. IEEE J. Biomed. Heal. Informatics. **20**, 1216–1224 (2016). https://doi.org/10.1109/JBHI.2016.2594058
52. Keum, D.H., Kim, S.K., Koo, J., Lee, G.H., Jeon, C., Mok, J.W., Mun, B.H., Lee, K.J., Kamrani, E., Joo, C.K., Shin, S., Sim, J.Y., Myung, D., Yun, S.H., Bao, Z., Hahn, S.K.: Wireless smart contact lens for diabetic diagnosis and therapy. Sci. Adv. **6**, 1–13 (2020). https://doi.org/10.1126/sciadv.aba3252
53. Elsherif, M., Moreddu, R., Alam, F., Salih, A.E., Ahmed, I., Butt, H.: Wearable smart contact lenses for continual glucose monitoring: a review. Front. Med. **9** (2022). https://doi.org/10.3389/fmed.2022.858784
54. Rabensteiner, D.F., Rabensteiner, J., Faschinger, C.: The influence of electromagnetic radiation on the measurement behaviour of the triggerfish® contact lens sensor. BMC Ophthalmol. **18**, 1–5 (2018). https://doi.org/10.1186/s12886-018-1013-x
55. Guo, S., Wu, K., Li, C., Wang, H., Sun, Z., Xi, D., Zhang, S., Ding, W., Zaghloul, M.E., Wang, C., Castro, F.A., Yang, D., Zhao, Y.: Integrated contact lens sensor system based on multifunctional ultrathin MoS_2 transistors. Matter **4**, 969–985 (2021). https://doi.org/10.1016/j.matt.2020.12.002
56. Aravind, M., Chidangil, S., George, S.D.: Self-moisturizing contact lens employing capillary flow. Addit. Manuf. **55**, 102842 (2022). https://doi.org/10.1016/j.addma.2022.102842

Chapter 3
A Spectrum of Choices: Finding the Right Contact Lens

The evolution of contact lenses over time has led to a diverse range of lenses tailored for various purposes. Contact lenses are ocular prosthetic devices utilized by over 150 million individuals globally [1]. The primary applications of contact lenses encompass eyesight correction, therapeutics, and cosmetic improvement. There are numerous choices available, ranging from hard to soft lenses, large to small sizes, and even colorless to highly colorful varieties, each catering to specific user needs. This chapter provides a thorough exploration of contact lenses, discussing their types, functions, and the significance of design and precision fit. It examines spherical, toric, and multifocal designs tailored to address various vision correction requirements. When selecting the right contact lens, it is crucial to consider factors beyond lifestyle and functions.

3.1 Types of Contact Lenses

3.1.1 Rigid Gas Permeable Contact Lenses

Hard contact lenses were the older version of rigid contact lenses introduced in the mid-twentieth century. These lenses, made from a hard, durable material called poly(methyl methacrylate) (PMMA), were admired for their optical clarity and ability to provide excellent vision correction [2]. These glass-like lenses, while offering excellent vision correction, were rigid and uncomfortable, requiring an extended adjustment period due to limited oxygen permeability [3]. The lack of oxygen flow to the cornea due to PMMA's Dk value of essentially zero can lead to various eye health issues, including corneal hypoxia, discomfort, and potential long-term complications. Since PMMA lenses do not allow oxygen to pass through the material, they rely entirely on tear exchange beneath the lens to supply the cornea with oxygen, which can be insufficient during extended wear [4]. To address these limitations,

© The Author(s), under exclusive license to Springer Nature Singapore Pte Ltd. 2025

N. Adrus et al., *Contact Lenses*, Series in BioEngineering,
https://doi.org/10.1007/978-981-96-3636-5_3

cellulose acetate butyrate (CAB) emerged as a base material in the late 1970s, representing a progression in hard contact lens technology [5, 6]. Commercial CAB lenses developed from Porofocon A (Table 3.1) offered slightly more flexibility and oxygen permeability compared to PMMA. Shortly after, butyl styrene (BS) lenses were introduced as another attempt to further enhance oxygen transmission beyond what was possible with PMMA and CAB [7]. While both CAB and BS lenses offered some degree of oxygen permeability compared to the PMMA, their performance in this area remained limited, especially when compared to the modern RGP lenses with significantly higher oxygen permeability. This difference in oxygen permeability has led to some authors categorizing CAB and BS lenses as traditional "hard" lenses [8], while others classify them as early forms of rigid gas permeable (RGP) lenses [9]. This distinction highlights the evolving understanding and classification of contact lens materials based on their ability to support corneal health through improved oxygen transmission.

Over time, the contact lens industry has evolved, transitioning from hard lenses to more advanced RGP materials. RGP lenses are a type of hard contact lens made from durable plastics that allow oxygen to pass through to the eye, ensuring corneal health and comfort [11]. Today's RGP lenses utilize advanced gas-permeable plastics, ensuring corneal health and comfort as shown in Table 3.1. The development of gas-permeable plastics revolutionized the contact lens industry, making RGP lenses the superior choice for those needing or preferring a rigid lens option. The table outlines different RGP materials categorized by their oxygen permeability (Dk) values. Low Dk siloxane/acrylate materials like Pasifocon A (Dk 14.6–15.6) and Lotifocon A (Dk 15.9) offer moderate oxygen transmission, making them suitable for patients requiring shorter wear times. In contrast, high Dk siloxane/acrylate materials such as Pasifocon C (Dk 56.0) and Telefocon B (Dk 43.5) provide greater oxygen transmission, which is beneficial for extended wear. Furthermore, fluoro-siloxane/acrylate materials like Tisilfocon A (Dk 250.0), Paflucon D, and Melafocon A (Dk 159.0) are designed for even higher oxygen permeability, combining the benefits of fluoro and siloxane components to enhance comfort and reduce the accumulation of deposits on the lens surface.

The advent of RGP lenses has largely made hard lenses with limited oxygen permeability obsolete, as they offer superior performance and user experience. RGP lenses continue to be an essential option in the contact lens market, catering to diverse eye conditions and individual needs. RGP lenses, made from fluorosilicone acrylate [12], fluoropolymers [13] and silicone acrylate [14, 15], represent a significant improvement in both comfort and oxygen transmission, crucial for maintaining eye health. Unlike PMMA lenses, RGPs allow oxygen to pass through the lens to the cornea, reducing the risk of hypoxia and other complications associated with reduced oxygen transmission [16].

The benefits of RGP lenses are manifold. They offer sharper vision compared to soft lenses, making them ideal for correcting astigmatism, keratoconus, and other irregular corneal conditions that soft lenses may struggle to address effectively [17]. Moreover, RGPs are more durable and cost-effective over their lifespan, despite a higher initial cost. This durability contributes to their popularity, especially in

Table 3.1 The commercial rigid gas permeable lenses, their source of materials and oxygen permeabilities (reused with permission from McMahon et al. [10])

Brand name	Material	Dk
Cellulose acetate butyrate (CAB)		
• *Meso*	Porofocon A	12.3
• *RX 56*	Porofocon A	12.5
Low Dk siloxane/acrylate		
• *Polycon II*	Silafocon A	12.0
• *Paraperm O2*	Pasifocon A	15.6
• *Boston II Pasifocon A*	Pasifocon A	14.6
• *Optacryl 60*	–	18.0
• *SGP-1*	Telefocon A	12.0
• *OP-2*	Lotifocon A	15.9
High Dk siloxane/acrylate		
• *Paraperm EW*	Pasifocon C	56.0
• *Boston IV*	Pasifocon B	28.7
• *SGP-2*	Telefocon B	43.5
• *SGP-3 43.5*	Unifocon A	43.5
Fluoro-siloxane/acrylate		
• *Menicon SFP*	Melafocon A	159.0
• *Menicon Z*	Tisilfocon A	250.0
• *Fluoroperm 30*	Paflucon C	30.0
• *Fluoroperm 60*	Paflucon B	60.0
• *Fluoroperm 92*	Paflucon A	92.0
• *Fluoroperm 151*	Paflucon D	151.0
• *Paragon HDS*	Paflucon B	58.0
• *Boston VII*	Satafocon A	73.0
• *Boston ES*	Enflucon A	36.0
• *Boston EO*	Enflucon B	82.0
• *Equalens*	Itafluorocon A	64.0
• *Fluorex 300*	Flusifocon C	30.0
• *Fluorex 500*	Flusifocon B	50.0
• *Fluorex 700*	Flusifocon A	70.0

orthokeratology [18] where lenses are worn overnight to reshape the cornea and improve vision temporarily. RGPs correct vision by using the tear film between the lens and the cornea to create a new refractive surface, essentially reshaping the cornea to correct vision.

While RGP lenses offer numerous advantages, it is important to acknowledge some drawbacks. Patients may initially find RGPs less comfortable than soft lenses, requiring an adaptation period with shorter wearing times recommended at first. Cost-wise, although RGPs have a longer lifespan [19], the initial cost is higher than soft lenses. However, the extended lifespan can potentially offset the initial investment over time. RGP users should be mindful of lens dislodgement during activities like contact sports. Opting for smaller diameter lenses can help to mitigate this [17]. Finally, RGPs necessitate specific cleaning and storage routines to maintain their hygiene and efficacy.

Despite these drawbacks, the RGP lens market continues to thrive due to ongoing advancements in lens technology and a diverse range of offers from leading manufacturers. Figure 3.1 illustrates several leading companies in the contact lens market that are recognized for their diverse range of RGP lens offerings. These companies, including BostonSight and Menicon, invest heavily in research and development, constantly refining materials, and designs to enhance oxygen permeability, comfort, and visual acuity. BostonSight excels in highly customized lenses for conditions like keratoconus, utilizing scleral lenses that provide exceptional comfort and stability on the sclera rather than the cornea [20]. Their Prosthetic Replacement of the Ocular Surface Ecosystem Treatment program (PROSE) further enhances care for severe dry eye and corneal disease patients. Beyond BostonSight's customized approach, Menicon stands out as a prominent manufacturer recognized for its diverse portfolio of RGP lenses, catering to a wide array of vision needs [21]. As highlighted in Table 3.1, Menicon utilizes advanced materials like Tisilfocon A and Melafocon A to achieve high oxygen permeability values in its lenses. For example, the Menicon Z Night, crafted from Tisilfocon A, boasts a remarkable Dk of 250, making it suitable for overnight wear in orthokeratology. Menicon Rose K for keratoconus with various geometries, and Menicon PremiO prioritizing comfort and eye health with high oxygen permeability and wettability [22], demonstrating their commitment to tailored vision solutions.

In addition to these leaders, Paragon Vision Sciences specializes in advanced RGP designs, providing tailored solutions for keratoconus and post-surgical vision correction. SynergEyes takes a different approach by innovating with hybrid lenses that combine the crisp optics of RGP centers with the enhanced comfort of soft lens skirts, addressing a range of vision needs. While known for soft lenses, Bausch

Fig. 3.1 RGP contact lenses representative major manufacturers according to market report

+ Lomb and CooperVision also offer customized RGP options for standard and complex prescriptions, focusing on comfort and precise vision correction. These offerings demonstrate the evolving and diverse capabilities within the RGP lens market, catering effectively to a broad spectrum of visual needs and eye conditions.

3.1.2 Soft Contact Lenses

Advancements in materials, particularly the development of soft contact lenses, have revolutionized vision correction. These soft lenses have provided a comfortable alternative to traditional eyeglasses and have been crucial in improving the vision and quality of life for millions of individuals [23]. The introduction of soft contact lenses was a significant breakthrough in the twentieth century, as they allowed for more comfortable and customizable designs [24].

Soft contact lenses are typically made from polymers that absorb large quantities of water, forming a hydrogel [25]. This makes them much more comfortable to wear than RGP lenses, especially for new wearers. The hydrogel material is also oxygen permeable, allowing the cornea to receive the necessary oxygen for maintaining healthy eyes [25]. Soft lenses are now an essential part of virtually every eyecare professional's toolkit for vision correction and a significant source of revenue [23]. Soft contact lenses are categorized by their water content and ionic charge, which affect their durability and resistance to protein deposits. The ionic charge refers to whether the lens surface is charged. Ionic lenses have a negatively charged surface, while non-ionic lenses have a neutral surface. This charge difference influences how the lens interacts with the tear film, affecting comfort and protein deposit formation. Ionic lenses are designed to attract water, making them easier to wet.

According to the Food and Drug Administration (FDA), soft lens materials are mainly categorized as hydrophilic (-filcon) orhydrophobic (-focon) based on their chemical composition (Table 3.2). For hydrophilic contact lens, it can be further divided into five groups [26, 27]. Group I lenses are non-ionic polymers with a lower water content (<50%), while Group II lenses are also non-ionic, with water content higher than 50%. Meanwhile, Group III and Group IV soft lens are ionic polymers, with Group III is having lower water content (< 50%) than Group IV (> 50%). Finally, Group V stands out by utilizing silicone hydrogel materials, which offer distinct advantages in oxygen.

In the case of hydrophobic lenses, their classification centers around the content of silicone and fluorine in the lens formulation, resulting in four distinct groups. These elements are crucial as they directly impact the lens's oxygen permeability and surface properties. Group I lens contain neither silicone nor fluorine and typically exhibit lower oxygen permeability. Group II lenses incorporate silicone, enhancing their oxygen permeability but lack fluorine. Group III lenses combine both silicone and fluorine, achieving high oxygen permeability and improved surface wettability due to the fluorine. Lastly, Group IV lenses contain fluorine for enhanced surface wettability but lack silicone, potentially resulting in "lower oxygen permeability."

Table 3.2 Classification of soft contact lenses according to US FDA (adapted with permission from González-Chomón et al. [28])

Classification	Group	Description
Hydrophilic	I	• Nonionic (< 1% ionized groups at pH 7.2) • Low water content (< 50%)
	II	• Nonionic (< 1% ionized groups at pH 7.2) • High water content (> 50%)
	III	• Ionic (> 1% ionized groups at pH 7.2), • Low water content (< 50%)
	IV	• Ionic (> 1% ionized groups at pH 7.2), • High water content (> 50%)
	V	• A new frontier in silicone hydrogel technology
Hydrophobic	I	• Without silicon and fluorine
	II	• With silicon but not fluorine
	III	• With silicon and fluorine
	IV	• With fluorine but not silicon

The origins of soft contact lenses can be traced back to the pioneering efforts of Otto Wichterle and Drahoslav Lím in the early 1950s [24, 25]. Their goal was to create a lens material that felt as natural as the eye itself while also providing clear vision. Initially, they explored polyvinyl alcohol for its use in human implant, but while it yielded optically clear hydrogels, it lacked the necessary mechanical strength for contact lenses. This setback highlighted a crucial principle in soft contact lens design: the need for shape stability and robust biochemical properties. For example, while transparency is essential, the polyvinyl alcohol (PVA) hydrogel lacked sufficient crosslinking, which is crucial for imparting mechanical strength and dimensional stability. Ultimately, Wichterle and Lím recognized the potential of poly (2-hydroxyethyl methacrylate) (pHEMA) for contact lenses. This hydrogel material offered comfort, ocular biocompatibility, and the ability to be molded into customized shapes.

By 1959, Wichterle's lab-scale PHEMA lenses were being tested on patients at the General University Hospital in Prague [29]. The soft contact lenses proved to be exceptionally well-suited for the eye, offering both comfortable wear and effective vision correction. The highly encouraging results from these pre-clinical trials prompted Wichterle to publish his views in Nature [30] the following year, outlining the key criteria for plastics in contact with biological tissues: transparency, biocompatibility, and sterilizability, which are essential for successful ophthalmic applications. Building upon his earlier work, Wichterle transitioned to prototype development, ingeniously creating a lens-making machine from a children's toy building set. This innovative spin-casting technique, enabling precise control over the lens shape, quickly demonstrated its potential for large-scale production: within the first quarter of 1962 alone, over five thousand lenses were manufactured. By 1971, their hard work paid off as pHEMA soft lenses became widely available, rapidly penetrating the market following FDA approval in 1972. This groundbreaking development forever changed the lives of millions, helping them see the world more clearly.

The introduction of pHEMA marked a turning point in the history of contact lenses, ushering in the era of "first-variation" soft lenses. This innovation was significant because pHEMA's ability to swell in both hydrophilic and hydrophobic environments, along with its excellent biocompatibility, made it a groundbreaking material for comfortable contact lenses [29]. These early soft lenses, including pHEMA copolymers incorporating *N*-vinyl pyrrolidone (NVP) and methyl methacrylate (MMA), offered unprecedented comfort and were initially affordable, leading to widespread adoption. However, their reliance on water channels for oxygen transport inherently limited their permeability, which is a crucial factor for corneal health [31]. While incorporating NVP enhanced oxygen permeability to a certain extent, and the addition of MMA further improved it, these modifications also introduced sensitivities to tonicity, pH, and heat. This spurred the development of second variation of soft contact lens, which incorporated silicone hydrogels as new materials. Silicone, known for its high oxygen permeability, allowed for the creation of lenses that could deliver significantly more oxygen to the cornea compared to their pHEMA-based predecessors. This advancement has been instrumental in enabling longer, healthier wearing times and reducing the risk of complications associated with oxygen deprivation. However, early silicone hydrogels sometimes faced challenges with surface wettability, impacting comfort for some wearers. Third-variation silicone hydrogels focused on improving surface technology to enhance wettability and comfort, aiming to mitigate the drawbacks of earlier silicone hydrogels while maintaining their superior oxygen permeability. Table 3.3 summarizes the evolution of soft contact lenses, highlighting key features, addressing challenges, and showcasing advancements in material science and lens design.

Table 3.3 The variations of soft contact lens materials

Materials used	Key features	Challenges	Advancements
pHEMA	Affordability and comfort	• Limited oxygen permeability • Reliance on water channels	• Incorporation of NVP for better oxygen permeability • Addition of MMA to improve permeability
Silicone hydrogels	• Higher oxygen transmission • Promoted corneal health	• Challenges with wettability • Potential discomfort	• Focus on enhancing oxygen permeability through silicone incorporation
Advanced silicone hydrogels	• Improved surface technology • Enhanced wettability	• Early silicone hydrogels had poor wettability	• Development of advanced surface coatings • Improved biocompatibility • Reduction in protein deposits and debris

Building on the foundational developments of first-, second-, and third-variation lenses, researchers are now focused on innovations that extend the functionality and environmental sustainability of soft contact lenses. Beyond third-variation lenses, innovations continue to redefine the field, particularly with research aimed at improving both environmental sustainability and functionality. Modern lenses now incorporate materials and coatings designed to be eco-friendlier, addressing concerns about the environmental impact of single-use lenses through biodegradable options. Additionally, the advent of 'smart' contact lenses introduces novel applications, such as lenses for continuous health monitoring, augmented reality displays, and controlled drug delivery, widening the scope beyond vision correction [32]. Alongside these advancements in material and functionality, the design of soft contact lenses has also been optimized for greater comfort and stability. Furthermore, the larger size of soft lenses contributes to their stability, making them less likely to dislodge during activities like sports, as shown in Fig. 3.2. This figure illustrates the difference between soft and rigid lenses on the eye, highlighting how the larger surface area of soft lenses enables them to conform closely to the eye's shape, providing improved stability and comfort. This stability, combined with their inherent comfort, makes soft lenses an excellent choice for individuals with active lifestyles or those who prioritize a comfortable wearing experience.

While soft contact lenses provide significant comfort advantages and enhanced stability, they also have distinct features that may influence their suitability for certain users. For example, soft lenses typically offer a shorter wearing time than rigid lenses, making them ideal for daily wearers who prioritize comfort and flexibility over extended wear. Additionally, soft lenses are more widely available for common vision corrections but may have limitations in achieving the highest levels of visual acuity or correcting specific conditions like corneal astigmatism. These lenses are also generally less durable than rigid lenses, as their softer material is less resistant to scratching or deformation.

Soft contact lenses provide notable comfort benefits and faster adaptation times, making them a popular choice for many users. Table 3.4 highlights the key features

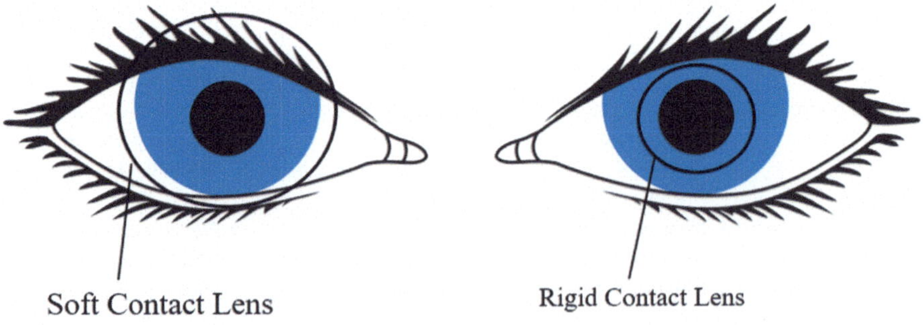

Fig. 3.2 Illustration of the key difference in diameter between soft and rigid contact lenses. The larger size of the soft lens enables it to fit comfortably under the eyelid, providing a more stable fit that is less likely to be dislodged

Table 3.4 Comparison between hard and soft contact lens

Characteristic	Hard contact lens	Soft contact lenses
Comfort	Less initial comfort provided to the wearers	Provides initial comfort to the wearer
Adaptation period	Required some time to adapt	Less time needed to adapt
Corneal astigmatism correction	Suitable for correcting corneal astigmatism	Not suitable for corneal astigmatism correction
Vision quality	Provide clear and crisp vision	Vision is less clear
Lens toughness	Can scratch and break	Does not usually deformed
Practicality	Not suitable for active movements involved with sports	Suitable for sports activities

that distinguish soft lenses from hard lenses, showing how each type meets different vision needs and lifestyle preferences. While soft lenses excel in comfort and flexibility, they generally have a shorter wearing time than hard lenses, often requiring more frequent removal. Additionally, not all optical powers are available in soft lens designs, which may limit options for some users. Soft lenses may also be less effective for certain vision issues, such as corneal astigmatism, and can sometimes offer lower visual acuity compared to hard lenses. Ultimately, choosing between soft and hard lenses depends on individual requirements. While soft lenses are well-suited for those prioritizing comfort, a thorough assessment of vision needs and consultation with an eye care professional can help identify the best option.

3.1.3 Hybrid Contact Lenses

Introduced in the 1980s, hybrid contact lenses (HCL) offer a unique and innovative solution for individuals seeking both sharp vision and comfortable wear. Take a look at this hybrid lens in Fig. 3.3 it is like the ultimate "do-it-all" contact lens! Imagine combining the high-definition optics of an RGP lens with the comfortable, soothing feel of a soft lens. This dynamic synergy is exactly what hybrid lenses accomplish! HCL offers the best of both worlds, providing precise vision without sacrificing comfort [33].

HCL presents a versatile solution for individuals who have faced challenges with the limitations of either soft or RGP lenses. The unique and innovative dual-material design is the key to their effectiveness: the sharp optics of a rigid gas-permeable center combined with the comfortable fit of a soft lens skirt (Fig. 3.4). The RGP center is commonly made of materials like PMMA enhanced with silicone or fluoropolymers to improve oxygen permeability. This modification is crucial for corneal health but does not directly contribute to sun protection. The RGP center primarily maintains the lens shape and corrects refractive errors for clear vision. Surrounding the RGP center is a soft lens peripheral material, typically made of silicone hydrogel. Silicone

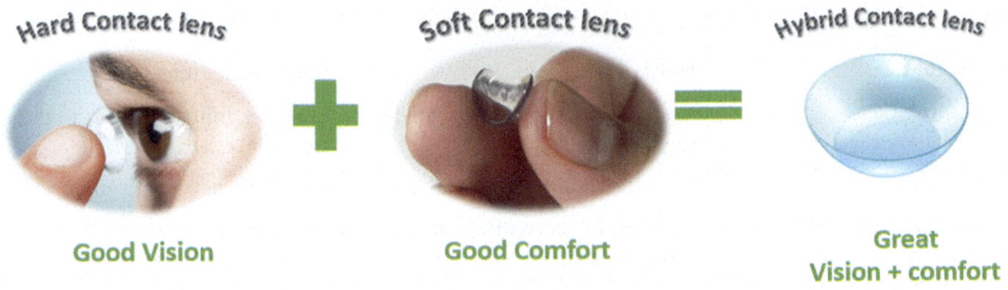

Fig. 3.3 Hybrid contact lenses combine the good vision of hard contact lenses with the good comfort of soft contact lenses to offer great vision and comfort

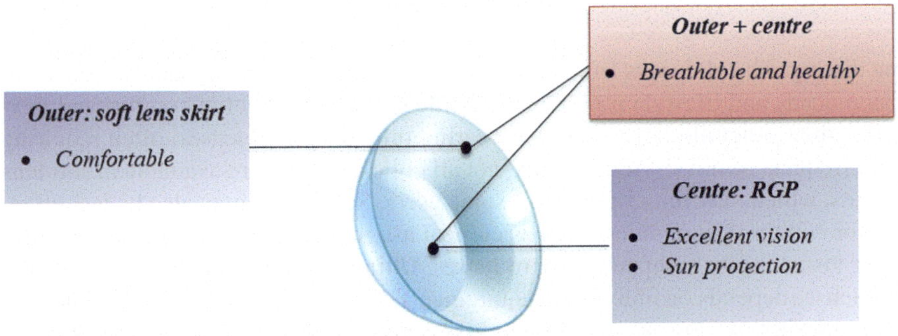

Fig. 3.4 Hybrid contact lens illustration. The rigid center provides excellent vision clarity, offering clear and crisp vision correction. The breathable and healthy design utilizes a patented dual material to enable the healthy flow of oxygen to the eye. The soft part of the lens ensures all-day comfort, and the inclusion of a UV blocker offers sun protection for the eyes

hydrogel is preferred due to its high oxygen permeability, which is essential for corneal health, especially during extended wear.

Table 3.5 summarizes a detailed comparison of the advantages and disadvantages of hybrid lenses in relation to RGP and soft contact lenses, highlighting key factors for wearers with normal and irregular corneas. Selecting the optimal lens type depends on both the anterior segment characteristics and the visual needs of each wearer. For both normal and irregular corneas, HCL provides the same optical performance as corneal RGP lenses, but with added benefits. When compared to corneal RGP lenses, hybrids are more comfortable, often exhibit better centration over the visual axis, and maintain greater stability. These advantages are particularly beneficial for irregular corneas, as the rigid portion of the hybrid lens is designed to vault over the corneal irregularity. Additionally, when compared to soft lenses, hybrids offer improved visual quality for patients with residual high-order aberrations or those for whom toric soft contact lenses fail to provide stable vision.

Individuals with specific visual needs, particularly those who have not found success with soft or rigid lenses, can experience significant benefits from the unique design of hybrid contact lenses. Studies have consistently demonstrated that HCL

Table 3.5 Pro and cons of hybrid contact lenses compared to RGP and soft contact lenses

	Hybrid versus RGP	Hybrid versus soft
Pros	• More comfortable • Quicker adaptation period • Easier to center the optics over the visual axis • More stable quality of vision • Vaulting results in less mechanical interaction with the cornea • Remains firmly positioned	• Better and more stable quality of vision • No need for lens stabilization to correct astigmatism • Better vision for patients who have significant high-order aberrations • More effective for correction of presbyopia in astigmatic patients
Cons	• More difficult to apply and remove • Longer time to settle • Must be replaced more often	• Higher cost • More difficult to fit • More careful evaluation of lens-to-anterior segment required • More difficult to apply and remove • Less frequent replacement

provides sharper vision, particularly for individuals with high astigmatism, presbyopia, or keratoconus (You can learn more about these vision irregularities in Sect. 3.2.1, Vision Correction. The RGP center of hybrid lenses excels at correcting astigmatism, while still maintaining a comfortable fit for the wearer. For individuals with presbyopia, the hybrid design can incorporate multifocal or bifocal optical zones, allowing for clear near, intermediate and distance vision. The larger diameter and customizable fit of hybrid lenses provide better lens centration and stability for the irregular corneas associated with keratoconus.

The development of HCL emerged from a long-standing effort within the field of optometry to reconcile the desirable visual acuity provided by rigid lenses with the comfort and stability offered by soft lenses. One of the earliest examples, the Saturn II lens, was a pioneering HCL that paved the way for the diverse range of designs available today. Second-generation lenses, like the SoftPerm lens (Sola/Barnes-Hind Incorporated) emerged in 1985, followed by the third generation, the SynergEyes lens (SynergEyes, Sola/Barnes-Hind Inc, Carlsbad, CA). These later designs addressed limitations of their predecessors, particularly in managing conditions like keratoconus, for instance offers both vault-driven designs and base curve-based designs for this purpose (SynergEyes lenses for keratoconus [34, 35]). In simpler terms, these designs differ in how the lens rests on the eye and distributes pressure fitting and available parameters. While each approach has pros and cons in terms of fitting and available parameters, they demonstrate the ongoing progress in HCL technology.

SynergEyes, a prominent manufacturer of HCL, offers a diverse portfolio of products, each design tailored to specific visual needs [36]. For instance, the SynergyEyes ID, available exclusively through eye care professionals, targets presbyopia, with or without astigmatism (Fig. 3.5). This lens incorporates a unique multifocal extended depth of focus design from the Brien Holden Vision Institute, a gas-permeable center, and a comfortable SiHy soft skirt. Another product, Duette, also available only through integrated eye Care Professionals (IECPs), is intended for a continuum

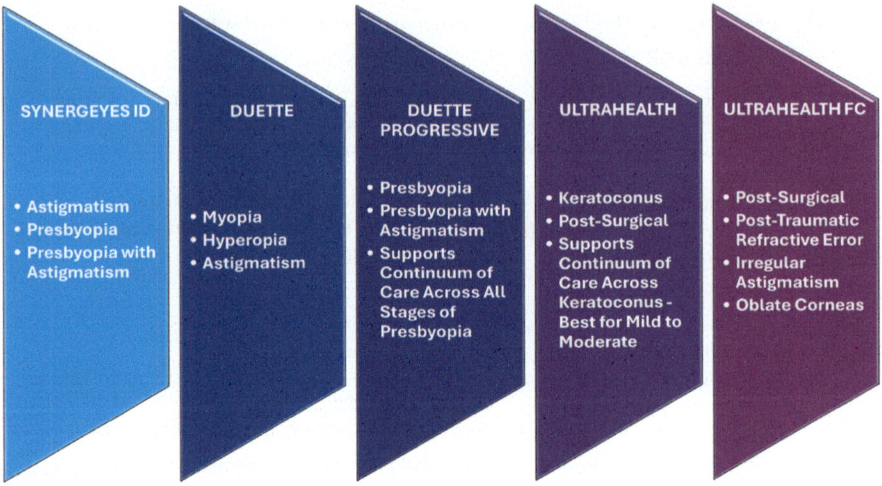

Fig. 3.5 SynergEyes portfolio of products

of care for keratoconus, particularly in mild to moderate cases. Its standout features include bi-tangential periphery, linear landing zones, and availability in Menicon Z and Contamac Optimum Extreme materials. Finally, UltraHealth lens caters to a wide range of post-surgical and irregular corneal conditions, including advanced keratoconus. This IECPs-exclusive lens features enhanced wettable surfaces, an ultra-thin edge, and UVA/UVB blocking for enhanced patient comfort and ocular protection.

HCL technology has progressed to its fourth and latest generation, with design like the Eyebrid® Silicone (Laboratoire LCS; France) and AirFlex® (SwissLens; Switzerland), incorporating aspheric designs with high Dk materials for improved oxygen permeability. These advancements along with a wider range of base curves and powers, allow for greater adaptability to a broader spectrum of visual needs.

For individuals seeking the clearest possible vision alongside enhanced comfort and stability, HCLs present a compelling solution. While patient-reported comfort levels are generally positive, some wearers may initially experience lens awareness, particularly a sensation of the lens edge on the eye. This sensation typically diminishes within a few days to a week as the eye adapts to the lens. Similarly, the visual experience is often described as sharper and clearer compared to soft lenses, especially for those with astigmatism. Some individuals may initially experience visual disturbances like halos or glare in low-light conditions. These disturbances usually subside with adaptation. The stable fit of hybrid lenses, characterized by good centration and minimal rotation, is another advantage. However, hybrid lenses typically come at a higher cost when compared to conventional soft lenses. In summary, hybrid lenses offer a balance of the benefits of both rigid gas permeable and soft contact lenses, combining crisp optics with enhanced comfort. A thorough evaluation with an eye care professional is essential to determine if the advantages of hybrid lenses outweigh the potential drawbacks and higher cost for individual wearers.

3.2 Functions of Contact Lenses

Contact lenses, small yet powerful devices placed directly on the eye, have revolutionized vision correction. Their primary function is to address refractive errors, acting as miniature lenses that work in harmony with the eye's natural optics to focus light onto the retina and provide clear vision. Whether compensating for nearsightedness, farsightedness, astigmatism, or age-related presbyopia, contact lenses offer a precise and customizable solution for achieving visual clarity. Beyond their corrective role, contact lenses have also become popular for cosmetic purposes, enhancing, or altering eye color, and even serve therapeutic functions, such as protecting the cornea or delivering medications.

3.2.1 Vision Correction

Among the available options for vision correction, contact lenses have emerged as a particularly popular choice due to their unique and compelling combination of convenience, aesthetic versatility, and the ability to provide wearers with a clear, sharp, and comfortable, that allows them to fully engage with immersive visual experience to the greatest possible extent. Their seamless integration with the eye's movements allows for accurate light bending and focusing, making them an attractive solution for individuals seeking to enhance their visual acuity and overall quality of life. Over time, a variety of contact lenses has been developed to correct common refractive errors as outlined in Table 3.6 illustrating the specific correction method employed for each condition. Soft contact lenses, crafted from flexible, water-absorbing plastics, offer comfort and are well-suited for correcting myopia, hyperopia, and mild astigmatism. For those with higher-order astigmatism and keratoconus, RGP lenses provide a more durable option with sharper vision. In principle, contact lenses provide a unique corrective mechanism, by acting as the initial point of contact for light entering the eye to focus more accurately.

When the eye's shape prevents light from focusing directly on the retina, refractive errors and blurred vision occur. These refractive errors can arise from variations in eyeball length, corneal irregularities, or age-related changes to the lens's flexibility. For instance, myopia, or near-sightedness, where distant objects appear blurry, occurs when the eye's shape causes light to focus in front of the retina. It is corrected with spherical lenses with a negative diopter, which reduces the eye's refractive power, bringing the focal point onto the retina (Table 3.6). Myopia often starts among children between ages 8 and 15, and can also be managed with orthokeratology lenses, which temporarily reshape the cornea. Conversely, hyperopia, also known as farsightedness, occurs when light focuses behind the retina, due to the eye being shorter in axial length, causing near objects to appear blurry. Convex lenses with a positive diopter increase the eye's refractive power to focus near objects on the retina

Table 3.6 Vision correction of common refractive error by contact lenses

Refractive errors	Correction method
Normal Vision 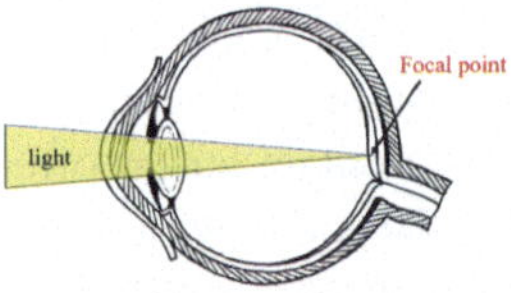	No refractive errors, as the focal point is formed directly on the retina. The focal point should be at this position after a specific lens design is applied to correct the refractive errors such as myopia, hyperopia, astigmatism, and presbyopia
Myopia 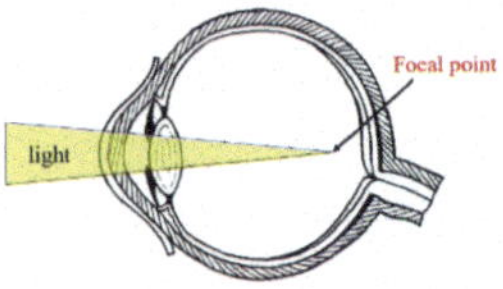	Use spherical lenses with a negative diopter to reduce the eye's refractive power, allowing distant objects to focus on the retina. Orthokeratology lenses are also an option for reshaping the cornea temporarily [37]
Hyperopia 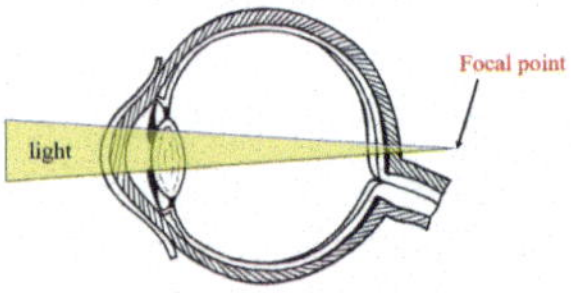	Use convergent or convex spherical lenses with a positive diopter to increase the eye's refractive power, helping near objects focus on the retina [38]
Astigmatism 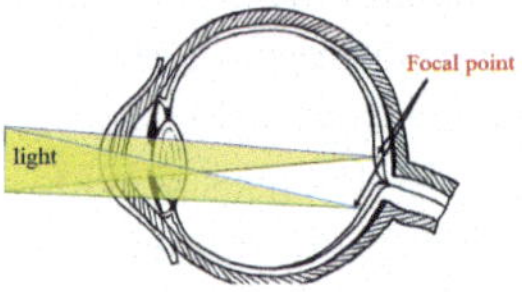	Regular astigmatism, caused by the cornea or crystalline lens, can be corrected using toric (cylindrical) lenses, which have a toric surface resembling a football with two different curves. Rigid gas permeable (RGP) contact lenses can also help by creating a tear lens that fills in corneal irregularities, providing a smooth optical surface [39]
Presbyopia 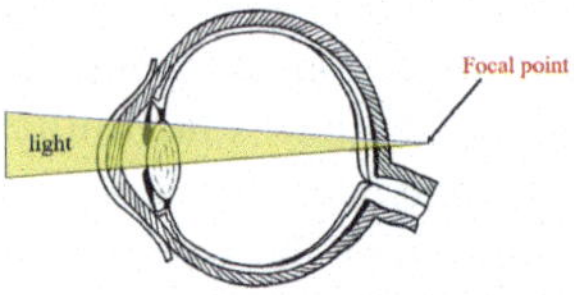	Use multifocal or bifocal lenses, which have multiple lens powers, to correct both distance and near vision. These lenses may have concentric zones or alternating segments to provide the necessary vision correction

(Table 3.6). Hyperopia can start since birth but often disappears during adolescence, depending on the individual.

Table 3.6 also shows "astigmatism" among other refractive errors, which can affect people of all ages until proper eye diagnosis is conducted. This condition is caused by an abnormal, oval-shaped cornea that distorts vision at all distances and requires toric (cylindrical) lenses, which offer different powers along perpendicular

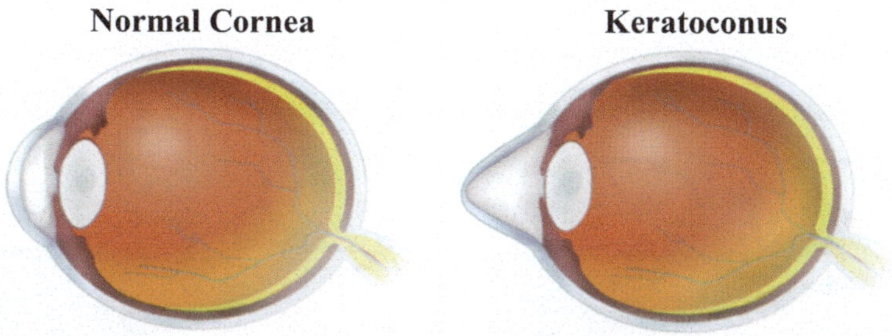

Fig. 3.6 Illustration comparing a normal cornea with keratoconus (reused with permission from [42] licensed under CC BY term)

axes to correct the irregular focus. RGP lenses also provide an effective solution for astigmatism, as their rigid structure creates a smooth optical surface by filling in corneal irregularities. Lastly, presbyopia —an age-related decline in near vision, typically occurring after age 40, as the lens of the eye hardens and loses flexibility over time. This condition is addressed with multifocal or bifocal lenses, which provide zones for near, intermediate, and distant vision correction [40, 41].

Beyond correcting common refractive errors, contact lenses can also address certain corneal irregularities, such as keratoconus, a condition where the cornea thins and bulges outward in a cone-like shape, as illustrated in Fig. 3.6. Specialized lenses, such as RGP lenses, scleral lenses, or hybrid lenses, are often used to correct keratoconus [16]. These lenses provide a smooth, uniform surface over the irregular cornea, allowing light to focus properly. Positioned directly on the eye's tear film, these lenses interact with incoming light rays, refracting the light to compensate for the distorted corneal shape and directing it accurately onto the retina. This corrected light path enables clearer vision, helping individuals with keratoconus achieve sharper and more stable vision than with standard soft lenses [7].

Unlike spectacles that sit at a distance from the eye, contact lenses move in unison with the eye's movements, providing a wider field of vision and eliminating obstructions caused by frames, as shown in Fig. 3.7. This feature is particularly beneficial for active lifestyles and sports, as contact lenses offer greater freedom of movement and stability. Table 3.7 outlines these advantages, showing the distinct benefits of contact lenses compared to spectacles.

From a vision standpoint, contact lenses often provide superior optical quality compared to spectacles. By placing it directly on the eye, contact lens eliminates distortions and aberrations that can occur when light travels through the air space between the eye and spectacle lenses. This direct contact allows for more precise correction of refractive errors, resulting in clearer and sharper vision. Additionally, contact lenses hold cosmetic flexibility, allowing for changes in eye color or subtle alterations to users' eye shape in ways that spectacles cannot. Furthermore, many individuals find contact lenses more comfortable than spectacles, as they eliminate the pressure and weight associated with frames resting on the nose and ears.

Fig. 3.7 Illustration comparing the field of vision provided by spectacles and contact lenses

Table 3.7 Advantages of contact lenses for vision correction

Advantage	Description
Wider field of vision	Contact lenses move with the eye, eliminating obstructions and reflections from frames, providing an unobstructed view and making them ideal for active lifestyles and sports
Greater freedom and stability	They offer greater freedom of movement compared to eyeglasses, as they do not slide or fall off during activities, providing stability for the wearer
Cosmetic benefits	Contact lenses allow for changes in eye color and subtle alterations to eye shape, offering cosmetic flexibility that eyeglasses do not provide
Comfort	Many find contact lenses more comfortable than eyeglasses, as they eliminate the pressure and weight from frames resting on the nose and ears
Superior optical quality	Contact lenses eliminate distortions and aberrations by positioning the corrective lens directly on the eye, resulting in clearer, sharper vision compared to eyeglasses
Personalized fit and correction	Professional fitting ensures that lenses match the individual's corneal shape, optimizing vision correction and comfort

In summary, Table 3.7 highlights the advantages of contact lenses, including a wider field of vision, comfort, stability, superior optical quality, and cosmetic benefits, all of which contribute to their suitability for various lifestyles and activities. However, it is important to acknowledge that some individuals may experience variations in visual performance with soft contact lenses compared to spectacles, potentially due to factors like changes in corneal physiology, ocular aberrations, and tear film structure associated with lens wear [43, 44].

In essence, contact lenses offer a highly effective and versatile solution for correcting a wide range of vision problems, from common refractive errors to complex corneal irregularities. Their direct interaction with the eye's optical system, coupled with their ability to move naturally with the eye, provides numerous advantages over traditional eyeglasses, including a wider field of vision, enhanced comfort, and greater freedom of movement. However, the unique considerations of contact lens fitting, the potential for variations in visual performance, and the importance of proper care underscore the need for professional guidance. Consulting with an eye care professional is paramount to ensure a safe, comfortable, and successful contact lens experience tailored to individual needs and lifestyles. Through personalized care and ongoing management, eye care professionals play a vital role in helping millions worldwide experience the transformative benefits of contact.

3.2.2 Cosmetic Appearance

Beyond their vision-correcting capabilities, contact lenses have also gained widespread popularity for cosmetic applications, allowing individuals to enhance or transform their eye color and appearance. Decorative contact lenses, also known as cosmetic or fashion lenses, come in a wide array of colors, patterns, and effects to suit various styles and occasions. Figure 3.8 provides a visual representation of the different types of decorative contact lenses such as opaque tinted lenses and color circle lenses.

Colored Lenses come in a dazzling array of options, available with or without vision correction. They can subtly enhance natural eye color or provide a complete transformation. Opaque tinted lenses are ideal for those with naturally dark brown eyes, allowing them to effortlessly adopt vibrant shades like blue, emerald green, or warm hazel. Enhancement tints work best for lighter eyes, intensifying their natural hue. The primary benefit of colored lenses is aesthetic enhancement, enabling users to express their personal style and elevate their natural features.

Patterned Lenses: Popular for themed events, costume parties, or theatrical performances, patterned lenses are designed to create dramatic and unique effects. These lenses, which are especially popular during festive seasons, are a go-to option for creating the perfect Halloween costume or cosplay look. Options include designs that mimic cat eyes, zombie eyes, and even specialized patterns inspired by pop culture or favorite idols, enabling wearers to imitate the appearance of beloved characters or celebrities. Often available in various colors, sizes, and attractive designs, patterned

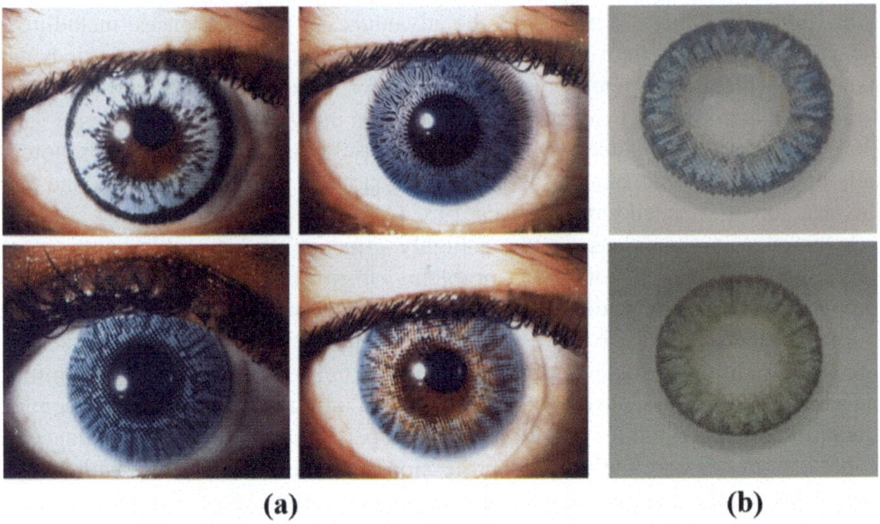

Fig. 3.8 Visual representation of the different types of decorative contact lenses such as **a** opaque tinted lenses (reused with permission from Fisher [45]), **b** circle lenses (courtesy from Alcon Johor)

lenses provide a fun and bold accessory to elevate costumes, add flair, and enhance personal style [26].

Circle Lenses: Originating in Asia, circle lenses have gained immense popularity for their ability to create a larger, doll-like eye appearance. These lenses extend the colored portion beyond the natural iris, resulting in a captivating, wide-eyed look that many find appealing. This effect is especially popular among individuals who enjoy experimenting with unique aesthetics or who are inspired by anime and doll-like features.

Decorative contact lenses have surged in popularity, providing a fun, accessible way to transform one's look, whether for a day, a special event, or an ongoing style choice. These lenses cater to a wide range of preferences and occasions, from enhancing natural eye color to creating dramatic effects with costume or theatrical lenses. For those who want to make a bold statement, the options are as limitless as the imagination from feline-inspired cat eyes and mesmerizing animal pupils to special effects lenses that glow in the dark or mimic otherworldly patterns. Whether it's for a costume party, cosplay event, theatrical performance, or simply to embrace a bold and adventurous aesthetic, decorative contact lenses offer a versatile playground for self-expression, allowing wearers to explore different facets of their style and identity.

3.2.3 Therapeutic Applications

Therapeutic contact lenses (TCL) have evolved beyond their original function in vision correction, now serving as innovative tools with far-reaching applications in modern medicine [6]. Traditionally, contact lenses were used solely for refractive error correction, but advancements in material science and design have enabled these lenses to play a significant role in managing ocular conditions. TCLs are increasingly valuable for therapeutic interventions grow significantly, with estimates suggesting a compound annual growth rate (CAGR) of 7–10% by 2023, offering a more consistent and sustained approach over traditional treatments such as eye drops or invasive procedures [46, 47].

Essentially, any type of contact lens can serve a therapeutic purpose, particularly when designed as a larger lens that blankets the cornea, creating a moist andprotective environment conducive to healing. Often referred to as "bandage lenses," these lenses extend beyond vision correction, serving as a vital tool in managing corneal and anterior eye diseases. TCLs are particularly beneficial in treating various ocular surface disorders including dry eye syndrome—a common condition where insufficient lubrication leads to irritation and blurred vision [48, 49]. TCLs can provide optimal hydration by supplementing the tear film and creating a moist environment, which is achieved when equilibrium is reached [48–50].

The mechanisms of both traditional and modern TCLs for treating dry eyes are illustrated in Fig. 3.9. As shown, TCL materials can contain embedded microparticles or thin films of medication, such as anti-inflammatory or lubricating agents. These particles gradually release the medication over time through diffusion or degradation, providing a steady delivery into the tear film and ensuring continuous relief while reducing inflammation without the need for frequent eye drop administration. For instance, lenses containing corticosteroids or cyclosporine are specifically designed to treat dry eye caused by inflammation [51, 52].

In addition to treating dry eye, TCLs are also utilized in managing blepharitis, a condition marked by inflammation of the eyelids, often due to bacterial overgrowth. TCLs alleviate blepharitis by acting as a mechanical barrier between the cornea and environmental irritants, which helps in maintaining moisture in the corneal epithelium, promotes wound healing, and provides pain relief [53]. Moreover, TCLs are valuable in addressing neurotrophic keratitis, a serious condition involving corneal damage from loss of sensation, potentially leading to ulcers and vision loss. Figure 3.10 highlights affected areas in various ocular diseases, managed by TCLs. By acting as a protective barrier, these therapeutic bandage lenses shield the delicate tissues of the anterior segment, promoting healing in damaged areas and preserving corneal moisture. Beyond these protective functions, TCLs are also being investigated for their potential intissue regeneration, as their biomaterial properties can be engineered to promote cell growth and facilitate tissue repair.

TCLs' minimally invasive nature, coupled with the potential for sustained and targeted drug delivery, has positioned them as a significant area of research and

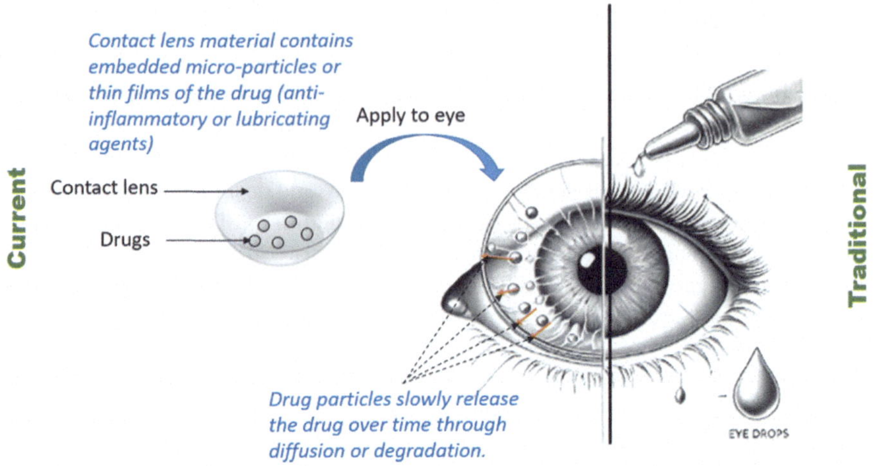

Fig. 3.9 Traditional and current treatment for dry eye. For current treatment, contact lenses have drug-delivery mechanisms, releasing anti-inflammatory or lubricating agents gradually to treat the underlying causes of dry eye

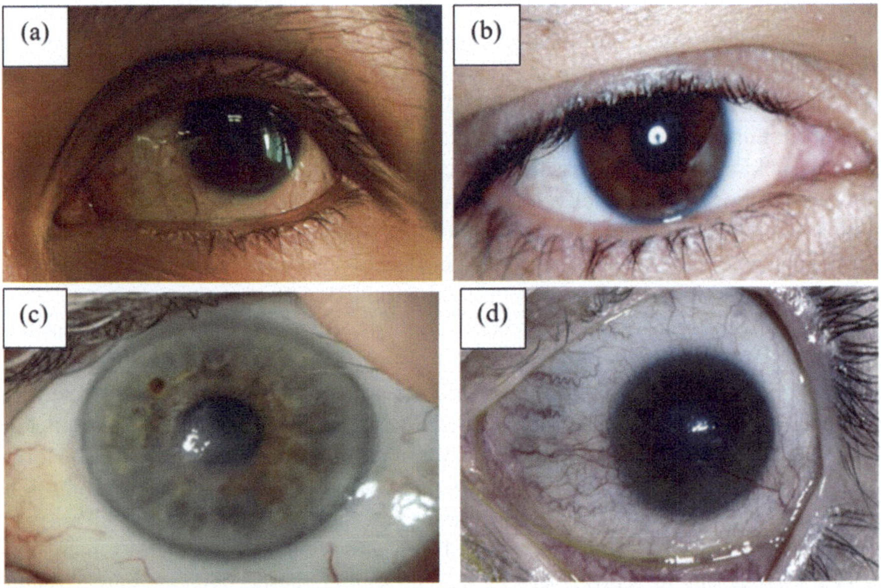

Fig. 3.10 Examples of ocular surface diseases, **a** dry eyes syndrome **b** Blepharitis (reused with permission from Bernardes et al. [54] **c** neurotrophic keratitis Blepharitis (reused with permission from Koay and Larkinn [55] licensed under CC BY) and **d** ocular rosacea (reused with permission from Vieira and Mannis [56])

clinical practice, particularly in ophthalmology. This shift towards therapeutic applications is driven by the growing significance of TCLs in addressing ocular and systemic conditions. Their targeted approach enhances efficacy while minimizing systemic side effects, making them a game-changer for managing conditions like glaucoma, dry eyes and even ocular infections [57]. Beyond ophthalmology, research is exploring their potential in diverse areas, including pain management, cancer treatment, and monitoring vital signs, highlighting TCLs' versatility and potential to revolutionize patient care across various medical specialties.

The advantages of therapeutic lenses are summarized in Table 3.8. These lenses are well-tolerated by patients, offering a comfortable pain-relieving shield for the eye. This is particularly beneficial following refractive surgery, where they provide crucial protection during the healing process. Beyond pain relief, therapeutic lenses offer a range of benefits, including promoting corneal healing by protecting newly formed cells, providing mechanical protection and support against injuries and diseases, maintaining ocular surface hydration in cases of severe dry eye, and enabling efficient drug delivery for improved permeation and absorption.

The selection of TCL is a highly individualized process. The ideal lens type depends on a patient's specific eye condition, the anticipated duration of treatment, and the unique needs of their cornea. To ensure optimal outcomes, a thorough eye examination and professional prescription from a qualified eye care professional are essential. This process extends beyond basic considerations of lens material and design covered in Chap. 6. A typical TCL prescription process involves:

- Comprehensive Eye Examination: This may include measuring corneal curvature, assessing tear production, and evaluating the overall health of the eye.

Table 3.8 Functions of therapeutic contact lenses

Aims/benefit	Description
Relief of pain	Increase comfort and pain relief from exposed nerve endings that can occur in conditions such as band keratopathy, corneal abrasions, and bullous keratopathy
Promotion of corneal healing	Assists healing of epithelial defects by protecting migrated and/or newly formed cells from the blinking action of the eyelids
Mechanical protection and support	Acts as a barrier in protecting the cornea in cicatrizing ocular surface diseases, and from mechanical injury in conditions such as trichiasis (inverted eyelashes abrading the anterior eye)
Maintenance of ocular surface hydration	Prevents tear evaporation or provides a moisture reservoir for the ocular surface in cases of severe dry eye
Efficient drug delivery	Allows prolonged drug delivery, albeit at a lower dosage rate, for better permeation and absorption because the drug remains in the eye longer

- Condition Assessment: The eye care professional will carefully assess the specific condition being treated, such as dry eyes, corneal abrasions, or post-surgical healing.
- Lens Selection: Based on the examination findings and the patient's needs, the most appropriate lens material and design are chosen. For example, a scleral lens might be selected for moisture retention, while a soft bandage lens could be prescribed for corneal protection.

This meticulous, patient-centered approach ensures that the TCL is tailored to the individual's specific requirements, ultimately enhancing both the effectiveness of the treatment and the patient's comfort.

Figure 3.11 presents an essential algorithm that outlines the key factors to consider before prescribing TCLs. This algorithm emphasizes the importance of evaluating both the potential advantages and contraindications, such as infective keratitis, corneal anesthesia, and severe exposure keratopathy. Additionally, extended wear of TCL carries an elevated risk of complications, including a greater incidence of infective keratitis. Therefore, prophylactic antibiotics and vigilant monitoring for complications like corneal hypoxia, inflammation, and allergic conjunctivitis are critical in TCL management [58].

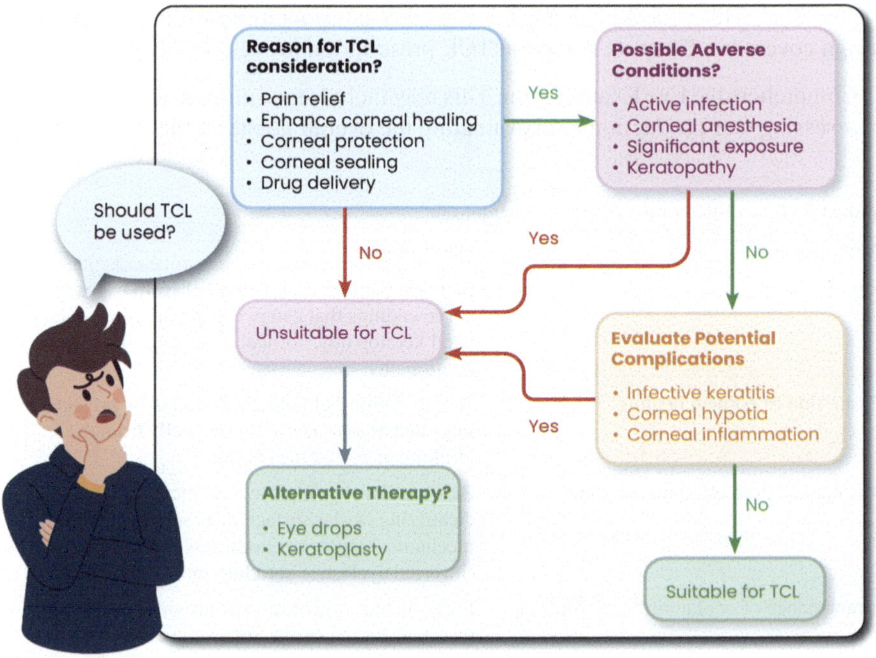

Fig. 3.11 Algorithm of therapeutic contact lens fitting, illustrating the decision-making process based on clinical considerations

TCL are available in a variety of materials to meet diverse therapeutic needs. Hydrogel lenses are soft and comfortable, though their oxygen permeability can vary. RGP lenses offer excellent oxygen transmission but may initially feel less comfortable. Scleral lenses, with their large diameter and corneal vaulting design, provide superior comfort and stability, making them ideal for complex corneal conditions. Silicone rubber lenses offer high oxygen permeability but can be more susceptible to deposits. Lastly, collagen shields, composed of natural collagen, serve as temporary lenses that dissolve over time, primarily used for short-term drug delivery and corneal protection [58, 59].

TCLs have redefined eye care, transcending from traditional vision correction to sophisticated means of treating various ocular conditions. With capabilities like drug delivery, corneal healing support, and protection of vulnerable ocular tissues, their impact extends beyond symptom relief, offering significant improvements in patients' quality of life. This progress, driven by continuous innovation in materials science and a deeper understanding of ocular physiology, has expanded TCL options from highly breathable silicone hydrogels to customized scleral lenses for customized patient care. However, their effective use requires meticulous patient evaluation, careful lens selection, and vigilant monitoring to mitigate complications.

Looking ahead, we can anticipate even more sophisticated TCL designs with enhanced biocompatibility, drug delivery capabilities, and therapeutic efficacy. Moreover, TCLs are expanding beyond traditional ophthalmology, with promising avenues emerging in areas such as continuous health monitoring, glucose tracking, and systemic diseases treatment. Overall, TCLs stand at the forefront of personalized ocular therapeutics, embodying a powerful blend of clinical need and scientific innovation.

3.3 Design and Precision Fit for Optimal Vision

In contact lens technology, achieving optimal vision is the primary goal. However, reaching clear, comfortable sight involves more than just determining the appropriate corrective power, it requires a careful understanding of how lens design and precision fitting work together to enhance visual acuity while safeguarding long-term ocular health. This section explores contact lens design, exploring the nuances of various parameters and their impact on visual performance. Three main types of contact lenses, each with unique design features and fitting considerations, will be explored: spherical lenses for simple refractive errors like nearsightedness and farsightedness, toric lenses for correcting the irregular corneal curvature seen in astigmatism, and multifocal lenses designed to address the age-related focusing challenges of presbyopia. The crucial role of individualized fitting protocols for each lens type will also be covered, highlighting how a precise fit ensures not only optimal vision but also comfort and the long-term well-being of the eye.

3.3.1 Spherical Contact Lens

Spherical lenses, the most common type of contact lens, are designed as a segment cut of a perfect sphere, with a uniform curvature across their entire surface. When a circle is rotated around an axis passing through its center, it forms a three-dimensional spherical surface. This geometry ensures that any radial cross-section of the lens, including the lens center or meridian, will have the same lens curvature and optical power curve setting spherical lenses apart from more complex designs such as toric or multifocal lenses. The distance between the sphere's center and its surface known as the radius of curvature is consistent across all points on the lens. A smaller radius translates to a steeper, more curved lens, while a larger radius indicates a flatter lens (Fig. 3.12, for further description on the dimension of contact lens, please refer to Sect. 4.2). The radius of curvature, also known as the base curve, is paramount in contact lens fitting as it must complement the cornea's shape to ensure optimal comfort, vision, and eye health. A steeper base curve (smaller radius) creates a tighter fit on the cornea, while a flatter base curve (larger radius) sits more loosely. Inadequate fitting can lead to discomfort, blurred vision, or even complications like corneal hypoxia, highlighting the need for precise measurements and manufacturing.

Spherical contact lenses are commonly recommended for patients with myopia or hyperopia who have minimal astigmatism, typically less than -0.75D. These lenses are available in daily, bi-weekly, and monthly replacement options. Power levels for spherical lenses are offered in 0.25D increments, increasing to 0.50D steps for powers above ± 6.00D, though specific increments may vary by brand. Determining the appropriate spherical lens power involves considering both the spherical equivalent and vertex conversion. The spherical equivalent provides an estimation of the prescription without astigmatism correction. It is calculated by halving the cylinder power from the eyeglass prescription and adding it to the spherical power.

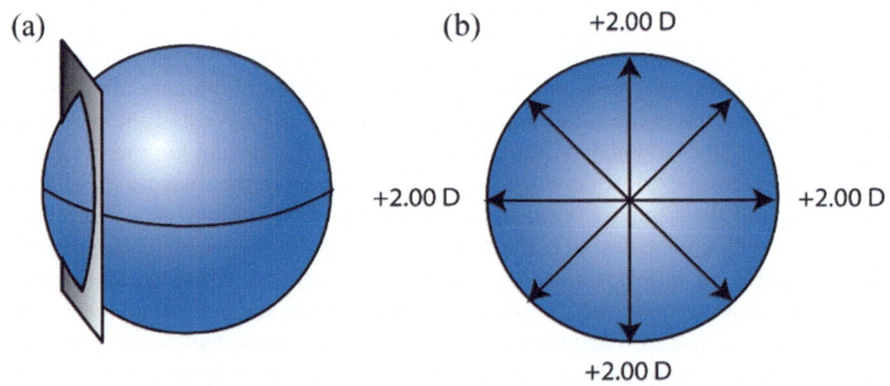

Fig. 3.12 **a** Illustration of spherical lens shapes and their corresponding radii of curvature. **b** A spherical lens provides equal power across all meridians

For instance, for an eyeglass prescription of + 2.00 – 0.50 × 180 would have a spherical equivalent of + 1.75D. When the cylinder power does not align with a multiple of 0.25, the practitioners use their professional judgment to select the closest spherical power.

For prescriptions with a power of ± 4.00D or higher, vertex conversion is necessary to account for changes in perceived lens power due to lens's distance from the eye. Eyeglass lenses typically have a vertex distance of 12 mm, while contact lenses sit directly on the eye (0 mm vertex distance). Using a vertex conversion chart (Table 3.9), practitioners determine the contact lens power equivalent based on the eyeglass prescription power. For example, an eyeglass prescription of – 8.00D would translate to a contact lens power of – 7.25D, which would then be rounded to the nearest available power, typically – 7.00D.

Figure 3.13 illustrates the working principles of spherical contact lenses, showing how convex and concave lenses adjust the focus of light onto the retina. These lenses operate on the principle of refraction, where light rays bend as they pass from the air into the lens material due to differences in density between the two mediums. The curvature of lens surface determines whether these light rays converge or diverge, to achieve the desired focus on the retina. Convex lenses, which are thicker at the center, cause light rays to converge, bringing the focal point closer to the retina and thereby correcting farsightedness (hyperopia). Conversely, concave lenses, which are thinner

Table 3.9 Back vertex power (BVP) sphere conversion table from CooperVision company, showing sphere values for lens prescriptions

Spheres	Minus	Plus	Spheres	Minus	Spheres	Minus
4.00	– 3.75	+ 4.25	7.75	– 7.00	11.50	– 10.00
4.25	– 4.00	+ 4.50	8.00	– 7.00	11.75	– 10.00
4.50	– 4.25	+ 4.75	8.25	– 7.50	12.00	– 10.50
4.75	– 4.50	+ 5.00	8.75	– 7.50	12.25	– 10.50
5.00	– 5.00	+ 5.50	9.00	– 8.00	12.50	– 11.00
5.25	– 5.25	+ 5.75	9.25	– 8.50	12.75	– 11.00
5.50	– 5.50	+ 6.00	9.50	– 8.50	13.00	– 11.50
5.75	– 5.75	+ 6.50	10.00	– 9.00	13.25	– 11.50
6.00	– 6.00	+ 6.75	10.00	– 9.00	13.50	– 11.50
6.25	– 6.25	+ 7.00	10.25	– 9.50	13.75	– 12.00
6.50	– 6.50	+ 7.50	10.50	– 9.50	14.00	– 12.00
6.75	– 6.75	+ 7.75	10.75	– 10.00	14.25	– 12.00
7.00	– 7.00	+ 8.00	11.00	– 10.00	14.50	– 12.00
7.25	– 7.50	+ 8.00	11.25	– 10.00		
7.50	– 8.00	+ 8.00	11.25	– 10.00		

The table provides the equivalent minus and plus corrections for different diopter values, based on Biofinity® conversion with a back vertex distance of 12 mm (adapted with permission from Huang [60])

at the center, cause light rays to diverge, pushing the focal point further back onto the retina lenses to correct near-sightedness (myopia).

Studies have also shown that spherical lenses can impact peripheral vision, particularly in managing myopia progression. Backhouse et al. [61] found that while traditional glasses often increase hyperopia (farsightedness) in peripheral vision, soft contact lenses generally reduce this effect. The study noted a significant shift toward myopic defocus in the periphery, especially on the temporal (outer) retina, which can be as high as 4 diopters in individuals with higher myopia. This level of peripheral myopia is even greater than that achieved by orthokeratology, a method known for its ability to slow myopia progression by inducing about 2 diopters of peripheral myopic defocus.

Moore et al. [62] further investigated the effects of specific spherical lens brands, Biofinity, Acuvue 2, Air Optix Night & Day Aqua, and PureVision2 on peripheral defocus across the nasal and temporal retina. As shown in Fig. 3.14, Biofinity, Acuvue 2, and Air Optix Night & Day Aqua generally induced a myopic shift (negative defocus) in peripheral vision, particularly on the temporal retina, where myopic defocus is associated with potentially slowing myopia progression. For instance, Biofinity and Air Optix Night & Day Aqua demonstrated significant myopic defocus at 30° and 40° in the temporal retina compared to the uncorrected eye, while Acuvue 2 showed a moderate myopic shift. In contrast, PureVision2's impact on peripheral defocus remained closer to the uncorrected baseline across both the nasal and temporal retina, indicating it may be effective for central vision correction without specifically targeting peripheral myopia control.

The differences in peripheral defocus across these brands may be influenced by their varying lens parameters, as shown in Table 3.10. While the base curves of these lenses vary slightly, no specific trend in peripheral defocus appears to correlate with base curve values. However, the center thickness does vary across brands, with PureVision2 having a slightly thinner center (0.07 mm) compared to the others (0.08 mm). This difference in thickness might contribute to PureVision2's minimal impact on peripheral defocus, as center thickness can influence the distribution of

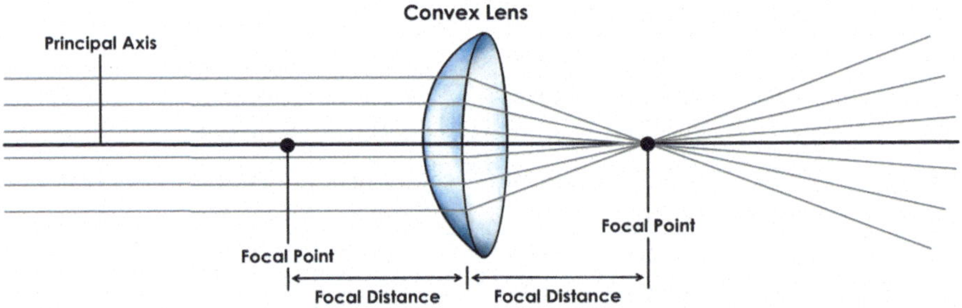

Fig. 3.13 Illustration of the working principle of a convex lens. Parallel light rays entering the convex lens are refracted and converge at a focal point after passing through the lens. The distance from the lens to the focal point is known as the focal distance. Convex lenses, thicker in the center, are typically used to correct farsightedness by focusing light onto the retina

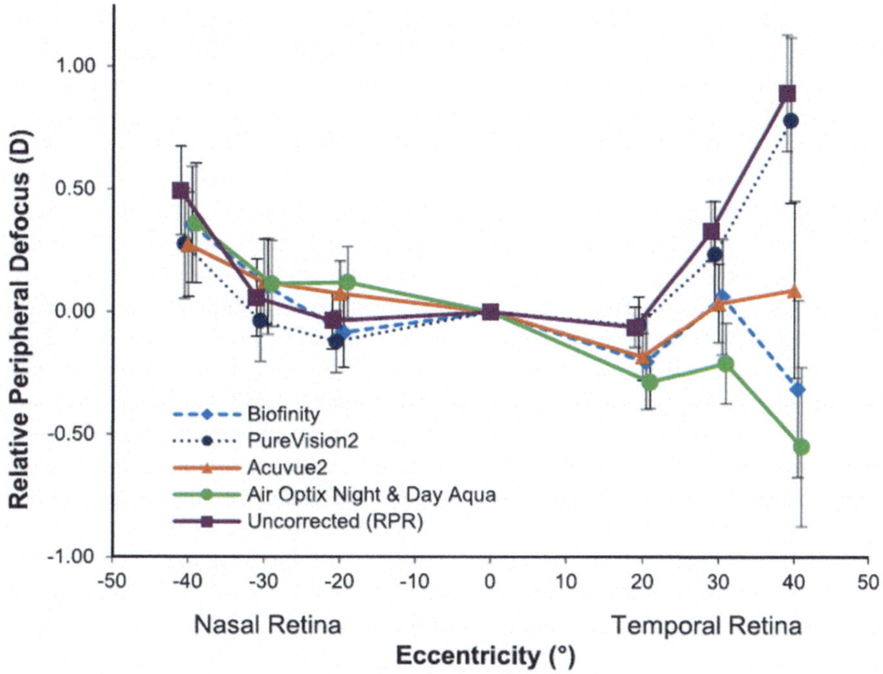

Fig. 3.14 Comparison of relative peripheral defocus (M) across the four soft contact lenses and in the uncorrected eye (no lens). Hyperopic defocus is shown by positive values, while myopic defocus is shown by negative values. Error bars indicate standard error of the mean (reused with permission from Moore et al. [62])

refractive power across the lens. Such variations highlight the importance of selecting lenses that induce peripheral myopic defocus when managing myopia progression is a priority [63].

Table 3.10 Specifications of contact lens parameters (mm) as reported by manufacturers for the lenses used in the study (reused with permissions from Moore et al. [62])

	Biofinity	Acuvue 2	PureVision2	Air Optix Night & Day Aqua
Material	Comifilcon	Etafilcon A	Balafilcon A	Lotrafilcon A
Base curve	8.6	8.3	8.6	8.4
Overall Diameter	14.0	14.0	14.0	13.8
Center Thickness (– 3.00OD)	0.08	0.08	0.07	0.08
Optik zone	*	*	9.0	8.0

*Not reported by the manufacturer

When selecting spherical lenses, users should consider both their vision needs and lifestyle. Spherical lenses not only correct common vision issues like nearsightedness andfarsightedness but also offer a convenient and discreet alternative to eyeglasses. They provide a wider field of view, reduce reflections, and well-suited for active lifestyles. Whether playing sports, engaging in hobbies, or simply enjoying the freedom of unhindered vision, spherical lenses seamlessly integrate into daily life. The choice of material and wear schedule often depends on individual preferences, lifestyle, and eye health considerations, with guidance from an eye care professional.

The development of silicone hydrogel, a material with high oxygen permeability, represents a significant. This material. Furthermore, as a versatile choice, spherical lenses offer significant customization, allowing them to be precisely tailored based on individual corneal measurements. This adaptability ensures a comfortable and accurate fit, even for users with slightly irregular corneas. Additionally, advancement in materials like silicone hydrogel, known for its high oxygen permeability, has further enhanced the performance of spherical lenses. By promoting corneal health and by reducing the risk of hypoxia, silicone hydrogel enables comfortable extended wear, making spherical lenses a reliable choice for many users. Finally, with the integration of digital tools, optometrists can now leverage advanced measurement systems to accurately assess the eye and recommend the most suitable spherical lens fit, ensuring both visual clarity and comfort for the wearer.

3.3.2 Toric Contact Lens: Mastering Astigmatism with Stability

Toric contact lenses are specially designed to correct vision problems associated with astigmatism; a common refractive error caused by an irregular curvature of the cornea. Unlike standard contact lenses, which feature a spherical shape resembling a slice of a beach ball, toric lenses have unique torus-like design that allows them to provide different refractive powers along the vertical and horizontal orientations. This specialized shape addresses the vision discrepancies associated with astigmatism, often resulting in clearer and more balanced vision for users [60]. Astigmatism often leads to blurred vision, difficulty seeing fine details, and distorted vertical lines. By providing targeted correction, toric lenses offer an effective solution for those who struggle with this common visual impairment.

Toric lenses come in various materials and formats to suit different preferences and lifestyles. They are available in daily disposable formats, soft materials, and in RGP lenses. While RGP lenses maintain their position on the eye more consistently and can provide sharper vision, they may require a longer adjustment period and are typically more fragile. Soft toric lenses offer prioritized comfort and ease of use, though they require precise fitting to stay properly aligned on the eye.

Correcting astigmatism with toric lenses has historically presented challenges, especially with soft contact lens, as residual astigmatism can reduce visual clarity and discomfort. First introduced clinically in 1979 [64], toric soft lenses were specifically developed to address full-eye astigmatism. Since then, advancements in design and fitting techniques have made them increasingly effective. Toric RGP lenses are often recommended for individuals with higher degrees of astigmatism because they maintain their shape on the eye and create a stable tear film beneath the lens, which allows for excellent optical performance. Nonetheless, varying levels of astigmatism among wearers can still impact both vision correction effectiveness and the success of lens fitting [64].

For users considering toric lenses, the degree of astigmatism plays a crucial role in determining the most suitable lens type (Table 3.11). Mild astigmatism (less than 1.00 D) may often be corrected with spherical soft or RGP lenses which may provide sufficient correction. For moderate astigmatism (1.00–4.00 D), toric soft or spherical RGP lenses are generally preferred for better alignment and stability on the eye. For higher level of astigmatism (over 4.00 D), toric RGP or custom soft toric lenses are recommended, to ensure optical performance and comfort. Careful selection based on these levels is crucial for achieve optimal visual acuity and comfort, meeting each wearer's specific needs effectively.

In addition to the degree of astigmatism, corneal topography can further guide lens selection. Reddy et al. developed a five-type classification system based on corneal toricity (Table 3.12), finding that an empirically based selection using this topography improved initial fitting success to 62%, with on-eye refinement raising success to 91%. This classification approach allowed for better alignment, particularly with back-surface toric designs like CooperVision Preference and Hydrasoft Torics, which showed stable fitting characteristics and high wearer comfort in eyes with varying toricity levels.

To enhance stability and prevent rotation during blinking, toric lenses incorporate specific stabilization features. Each stabilization method is designed to match the user's degree of astigmatism, promoting stable vision correction andpreventing misalignment [64]. Common methods include prism ballast, which adds extra weight to thicken the inferior zone of the lens (Fig. 3.15), and truncation, where the bottom edge of lens is flattened to help prevent rotation (Fig. 3.16). Soft toric lenses typically used either dynamic or a prismatic stabilization with beveled ballast or thin zones (Fig. 3.14), while RGP toric lenses often stabilize more effectively with a back toric surface. Research by Wolffsohn et al. [41] examined stabilization techniques for prism-ballasted, truncated toric lenses, revealing that lens diameter (13.5 × 12.5 mm vs. 12.8 × 11.8 mm) or prism diopters (0.75 vs. 1.25) had minimal effect

Table 3.11 Recommended lens selection based on the amount of astigmatism (reused with permissions from Filho et al. [65] licensed under CC BY 3.0)

Amount of astigmatism	First choice of lens
< 1.00 D	Spherical soft or RGP
1.00 to 4.00 D	Toric soft lens or spherical RGP
> 4.00 D Toric	RGP or custom soft toric lens

Table 3.12 Five-type classification system of corneal toricity

Type	Description	Central and peripheral toricity
Type I	Spherical cornea	$\leq$ 0.75 D central and peripheral toricity
Type II	Greater central toricity than peripheral toricity	Central toricity > peripheral toricity by $\geq$ 0.75 D
Type III	Limbal-to-limbal toricity (toricity extends across the cornea)	Central-to-peripheral toricity difference < 0.75 D
Type IV	Peripheral toricity exceeds central toricity	Peripheral toricity > central toricity by $\geq$ 1.00 D
Type V	Irregular peripheral toricity	Highly variable or irregular toricity across the periphery

on rotational stability. Their findings indicated that once the lens achieved stabilization, rotational movement remained under 5°, irrespective of the lens diameter or prism strength. They also found that steepening the base curve by 0.2 mm decreased rotational movement, indicating that a tighter fit improves stability.

A recent study by Gustafson et al. [67] assessed the fit and stability of the verofilcon A toric daily disposable lens. This silicone hydrogel lens with PRECISION BALANCE 8|4® stabilization demonstrated rapid settling-and excellent stability, orienting within an average of 3° from the ideal position within two minutes of insertion. The study reported that 98.7% of the lenses showed minimal oscillation ($\leq$ 5°) with blinking, highlighting the lens's stability and fit success. With a high-water content surface (SMARTSURFACE® Technology), this lens provided both stability and comfort, achieving a 99% first-fit success rate. These results make verofilcon an ideal choice for wearers with astigmatism seeking a predictable and easy fit.

In conclusion, selecting the appropriate toric lens design based on the wearer's degree of astigmatism and corneal characteristics is essential for achieving optimal visual clarity, stability, and comfort. Scientific advancements, particularly in

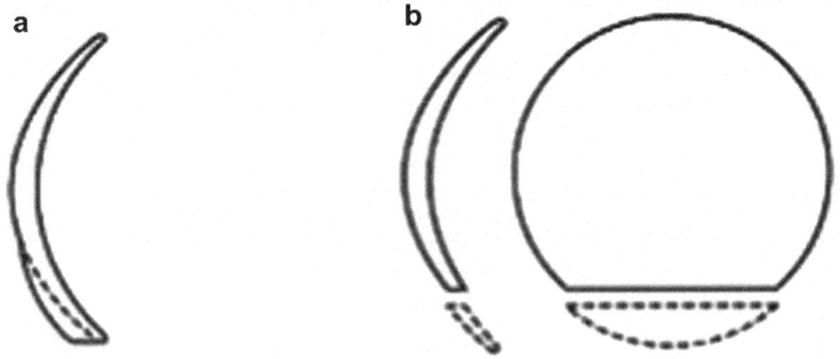

Fig. 3.15 Stabilization of toric contact lens **a** adding prism ballast, **b** flatten or truncation of the bottom edge of contact lens (reused with permissions from Filho et al. [65])

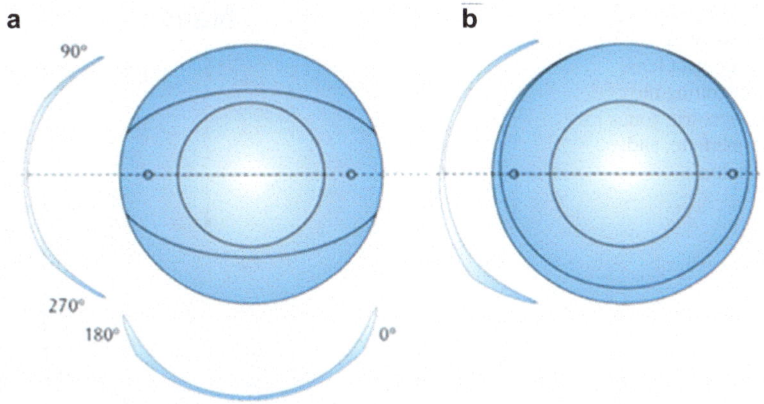

Fig. 3.16 Soft toric contact lens design **a** Dynamic toric stabilization system utilized in toric soft contact lens design, allowing for the correction of astigmatism up to 8.00 D. **b** Prismatic stabilization system positioned in the lower area of the contact lens (reused with permissions from Herranz et al. [66] licensed under CC BY 3.0)

topography-guided selection and empirical classification of corneal toricity, allow practitioners to customize lenses more effectively, improving initial fit success and long-term satisfaction. By considering stabilization methods, base curve adjustments, and material selection, eye care professionals can help users find lenses that align with their specific visual needs and lifestyle. Ultimately, toric lenses offer a reliable, personalized solution for astigmatism correction, allowing users to experience clear, stable vision across various daily activities.

3.3.3 Multifocal Contact Lens: Expanding Vision for Presbyopia

As presbyopia progresses, multifocal contact lenses provide an adaptable solution for users who need clear vision across various distances. The design and fitting of multifocal contact lenses require a precise balance between visual clarity and comfort across multiple focal distances. Soft multifocal simultaneous image contact lenses, particularly for their ability to support a seamless visual experience across distances, have been well-reviewed. Multifocal lenses, including bifocals, trifocals, or progressive lenses, enable continuous focus adjustment from near to far. These lenses incorporate distinct zones that allow presbyopic patients to seamlessly focus on near, intermediate, and distant objects. For instance, Fig. 3.17 shows concentric ring designs where the innermost ring supports near vision, the middle ring handles intermediate tasks, and the outer ring aids in distance clarity. Proper fitting ensures the user's eyes transition smoothly across these zones, allowing for easy adaptation to changing visual needs.

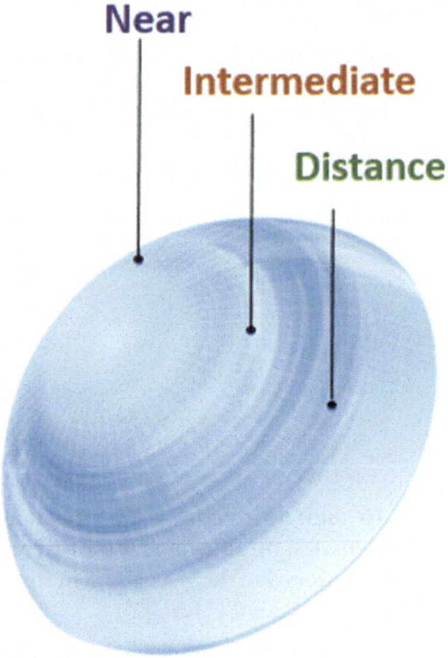

Fig. 3.17 Schematic representation of a multifocal contact lens design, illustrating the distinct concentric zones for near, intermediate, and distance vision

Multifocal lenses are typically recommended in later stages of presbyopia, especially as the add power required for near vision increases to around + 1.50 D or more. An overview of the available correction options for presbyopia is provided in Table 3.13, indicating that simpler methods like refractive compensation, or monovision are effective for the earlier stages. In standard monovision, one eye (usually the dominant eye) is corrected for distance vision and the other eye is corrected for near vision. As presbyopia advances, modified monovision or multifocal lenses become more beneficial. In modified monovision, for example, the dominant eye may use a monofocal or distance-focused multifocal lens, while the non-dominant eye wears a multifocal or trifocal lens covering near and intermediate vision. For advanced stages, where multifocal or monovision adjustments may be challenging, single-vision lenses with supplemental spectacles for specific tasks, like reading, can be effective.

Table 3.13 Strategies for correcting presbyopia with contact lenses across stages, including multifocal lenses for advanced needs (reused with permission from Schornack et al. [68])

Stage of presbyopia	Technique
Early (+ 0.75 add)	Refractive compensation
Early and moderate (+ 0.75 to + 1.50 add)	Monovision
Early, moderate, and advanced (any add)	Bifocal/multifocal contact lenses
Moderate and advanced (+ 1.50 to + 2.50 add)	Modified monovision
Individuals unable to adapt to the cited techniques	Single vision contact lenses, with supplemental spectacle correction

Multifocal lenses are available in two primary designs: alternating vision and simultaneous vision. Alternating vision lenses, also known as segmented multifocal lenses, are designed with distinct zones to help the eye shift focus based on vision needs, as depicted in Fig. 3.18. The top section of the lens is used for distance vision when looking straight ahead, while the lower section is dedicated to near vision when the gaze shifts downward. These lenses are often made with rigid gas permeable (RGP) materials and include stabilizing features, such as a ballast or a truncated bottom edge, to prevent rotation and ensure they remain aligned with the lower eyelid for stability and clarity [69].

In contrast, simultaneous vision lenses present images from various distances at once, requiring the brain to select the clearest image based on the visual target. Within simultaneous vision lenses, two main subtypes are commonly used: aspheric and concentric ring designs, also shown in Fig. 3.18. Aspheric designs feature a gradual change in refractive power from the center to the periphery, indicated with a color gradient. This gradual transition minimizes abrupt power changes, enhancing comfort and reducing visual disturbances like ghosting, allowing for a smoother shift between near and distant vision. Concentric ring designs, on the other hand, divide power into distinct circular zones, with alternating rings dedicated to different focal

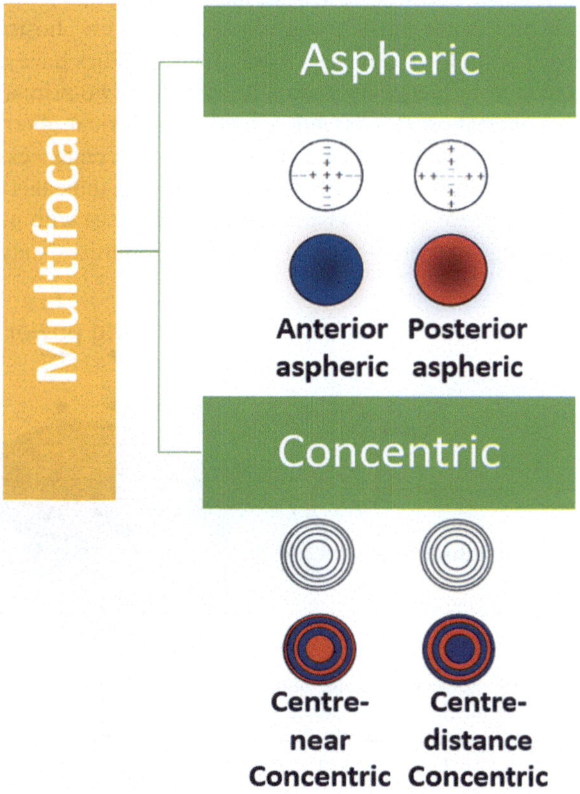

Fig. 3.18 Classification of multifocal contact lenses into aspheric and concentric designs (reused with permission from Remon et al [70] licensed under CC)

distances. Typically, the inner rings handle near vision, while the outer rings support intermediate and distance vision, accommodating various patient needs [70].

To assess the performance of multifocal contact lenses, wavefront-sensing technology is used to map power profiles across the lens surface. By analyzing light distortions as they pass through the lens, this profiling displays the distribution and magnitude of add power in multifocal lenses, helping practitioners in fine-tuning the lens's fit and functionality. In addition, Fig. 3.19 shows how power transitions smoothly from the center-near multifocal lens to the periphery facilitated by the lens's radial symmetry. Power distribution profiles provide insights into how various multifocal lenses are designed, revealing whether the lens transitions between zones are gradual or abrupt. Lenses with large, sudden power changes can cause visual disturbances, such as ghosting, due to the rapid shifts in focus. In contrast, smoother transitions found in aspheric designs enhance comfort, but may slightly limit the depth of field. Understanding these variations helps eye care professionals select lenses that align with a patient's visual needs, facilitating easier adaptation and maximizing visual performance across different distances.

Previous research on multifocal contact lenses has evaluated visual performance factors, particularly the effect on ghosting—a visual artifact where faint, unidentifiable images appear due to light scattering or misalignment. Kollbaum et al. [72] developed a Ghosting Questionnaire to quantify this effect, revealing that focus was the factor most strongly associated with ghosting ratings among presbyopic users of multifocal lenses. Additionally, studies have explored spherical aberrations and their influence on clarity, highlighting two primary aberrations: primary spherical aberration (PSA) and secondary spherical aberration (SSA). PSA fine-tunes clarity for near or distance vision, while SSA enhances overall sharpness and reduces ghosting. Research by Zheleznyak et al. [73] indicates that lenses with more negative PSA improve overall vision scores, and lenses with negative SSA further minimize ghosting. Although certain visual compromises may occur at specific distances,

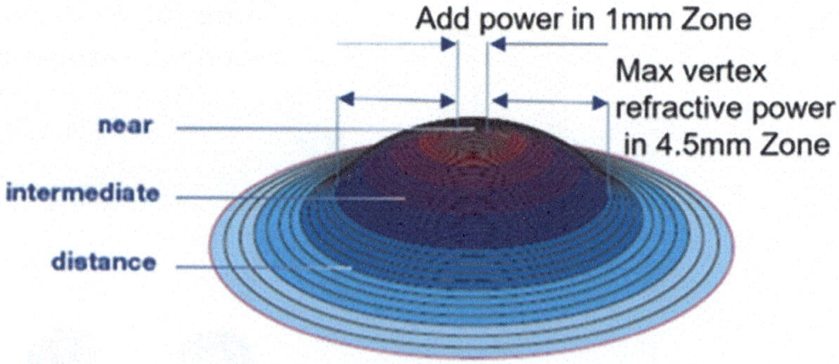

Fig. 3.19 Power profile of a multifocal contact lens showing the gradual transition of power across near, intermediate, and distance vision zones. The curve highlights how the lens manages vision clarity by smoothly adjusting power from the center to the periphery, ensuring balanced focus for different visual tasks (reused with permission from Kim et al. [71] licensed under CC BY)

lenses with negative PSA and positive SSA are generally preferred for offering balanced clarity and comfort across various visual needs.

Expanding on this foundation, Fedtke et al. [74] conducted a comprehensive study to assess the impact of PSA and SSA on visual performance in different multifocal contact lens designs. Seven lenses were tested, including a single-vision control (Air Optix Aqua SV), bifocal (Acuvue BF), center-distance (Proclear MFCL D), and center-near designs (e.g., Proclear MFCL N and Air Optix Aqua MFCL N). Table 3.14 organizes these lenses by presbyopia stage and user satisfaction, guiding lens selection based on presbyopia progression and visual needs. Results indicated that Acuvue BF caused the smallest PSA shift, while Proclear MFCL D induced positive PSA. All center-near lenses produced negative PSA, with the most pronounced shift in Proclear MFCL N and the smallest in Clariti 1 Day MFCL N. Visual performance varied across lenses: Air Optix Aqua SV excelled at distance vision, underscoring distance clarity compromises in multifocals, while multifocals generally outperformed the control at near vision. However, Acuvue BF and Proclear MFCL N, both with steeper power transitions, showed the lowest distance satisfaction. Air Optix Aqua MFCL N consistently reduced ghosting, achieving the highest satisfaction among presbyopic users.

In conclusion, multifocal contact lenses offer presbyopia users a convenient solution for clear, adaptable vision across near, intermediate, and far distances, helping to maintain comfort and visual clarity throughout the day. With advanced designs like simultaneous vision lenses, which provide smooth transitions through aspheric or concentric ring zones, and alternating vision lenses with distinct areas for varying distances, multifocal lenses cater to a wide range of visual needs. Power profiling technologies, such as wave front-sensing, further enhance these lenses by refining add power distribution, resulting in better adaptation and minimized visual disturbances. As presbyopia progresses, multifocal lenses provide a flexible, user-friendly option that supports dynamic vision needs, ensuring a seamless visual experience across multiple distances without compromising comfort.

Table 3.14 Recommendations for multifocal contact lenses based on presbyopia stage and visual performance satisfaction (adapted with permission from Fedtke et al. [74])

Presbyopia stage	Lens type	Lens brand	PSA/SSA characteristics	User satisfaction	Notes on visual performance
Early presbyopia	Single vision	Air Optix Aqua SV	Neutral PSA, SSA balanced	Low 5.4	Best distance performance, limited near vision; ideal for users needing supplementary spectacles for near tasks
Early presbyopia	Bifocal	Acuvue BF	Neutral PSA, Neutral SSA	Low 5.5	Suitable for early presbyopia, offering a gentle visual adjustment with minimal aberrations, though it may lack strong near correction
Early to moderate presbyopia	Center-distance MFCL	Proclear MFCL D	Positive PSA, Balanced SSA	Moderate 6.1	Clear distance vision; some compromise in near performance. Suitable for early-to-moderate presbyopes needing balanced distance and near vision
Moderate presbyopia	Center-near MFCL	Air Optix Aqua MFCL N	Negative PSA, Balanced SSA	Highest 7.0	Consistent performance across distances, minimal ghosting; good overall satisfaction for presbyopes with moderate near vision needs
Moderate to advanced presbyopia	Center-near MFCL	Clarity 1 Day MFCL N	Slight negative PSA, positive SSA	High 6.8	Stepped profile reduces abrupt transitions, providing smoother vision shift, suitable for patients needing moderate near correction and adaptability

(continued)

Table 3.14 (continued)

Presbyopia stage	Lens type	Lens brand	PSA/SSA characteristics	User satisfaction	Notes on visual performance
Advanced presbyopia	Center-near MFCL	Proclear MFCL N	High negative PSA, positive SSA	Lowest 4.8	Provides enhanced near vision; however, significant ghosting and visual disturbances reported, reducing satisfaction for advanced presbyopia users requiring near-dominant vision
Advanced presbyopia	Center-near MFCL	PureVision MFCL N	Negative PSA, positive SSA	Moderate 6.6	Good near performance, some ghosting at distance; suited for advanced presbyopia with emphasis on near vision but balanced adaptation needed

References

1. Moreddu, R., Vigolo, D., Yetisen, A.K.: Contact lens technology: from fundamentals to applications. Adv. Healthc. Mater. **8**, 1–24 (2019). https://doi.org/10.1002/adhm.201900368
2. Ridley, F.: Recent developments in the manufacture, fitting and prescription of contact lenses of regular shape. J. R. Soc. Med. **39**, 842–848 (1946). https://doi.org/10.1177/003591574603901227
3. Musgrave, C.S.A., Fang, F.: Contact lens materials: a materials science perspective. Materials (Basel) **12**, 1–35 (2019). https://doi.org/10.3390/ma12020261
4. Hill, R.M.: Four views of hypoxia ..., but which is best? Part 3. Using PMMA (Dk = 0) as a reference. Int. Contact Lens Clin. **27**, 22–23 (2000). https://doi.org/10.1016/S0892-8967(00)00040-7
5. Even-Shoshan, R., Gueichman, D.: The hard gas permeable lens as a first choice lens. J. Br. Contact Lens Assoc. **5**, 156–159 (1982). https://doi.org/10.1016/S0141-7037(82)80014-1
6. Ioniță, M., Vlăsceanu, G.M., Toader, A.G., Manole, M.: Advances in therapeutic contact lenses for the management of different ocular conditions. J. Pers. Med. **13** (2023). https://doi.org/10.3390/jpm13111571
7. Gurnani, B., Kaur, K.: Contact lenses. Br. Med. J. **288**, 1–16 (2023). https://doi.org/10.1136/bmj.288.6417.586
8. Yang, W.H., Smolen, V.F., Peppas, N.A.: Oxygen permeability coefficients of polymers for hard and soft contact lens applications. J. Memb. Sci. **9**, 53–67 (1981). https://doi.org/10.1016/S0376-7388(00)85117-0
9. Hales, L.: Sosa: Defining the Limits of the Maternal Sensitive Period. Dev. Med. Child Neurol. **19**, 454–461 (1977). https://doi.org/10.1111/j.1469-8749.1977.tb07938.x
10. McMahon, T.T., Zadnik, K.: Twenty-five years of contact lenses: the impact on the cornea and ophthalmic practice. Cornea **19**, 730–740 (2000)

11. Ichijima, H., Cavanagh, H.D.: How rigid gas-permeable lenses supply more oxygen to the cornea than silicone hydrogels: a new model. Eye Contact Lens **33**, 216–223 (2007). https://doi.org/10.1097/01.icl.0000254523.63767.05
12. Ren, L., Yin, S., Zhao, L., Wang, Y., Chen, H., Qu, J.: Study on the surface of fluorosilicone acrylate RGP contact lens treated by low-temperature nitrogen plasma. Appl. Surf. Sci. **255**, 473–476 (2008). https://doi.org/10.1016/j.apsusc.2008.06.060
13. Ramharack, R., Nguyen, T.H.: Fluoropolymers of very low surface energies. J. Polym. Sci. Part C Polym. Lett. **25**, 93–98 (1987). https://doi.org/10.1002/pol.1987.140250302
14. Collins, M.J., Brown, B., Bowman, K.J.: Contrast sensitivity with contact lens corrections for presbyopia. Ophthalmic Physiol. Opt. (1989). https://doi.org/10.1111/j.1475-1313.1989.tb00832.x
15. Lakkis, F.G., Cruet, E.N.: Cloning of Rat Interleukin-13 (IL-13) cDNA and analysis of IL-13 gene expression in experimental glomerulonephritis (1993)
16. Bennett, E.: Innovations in gas permeable multifocal contact lenses. Clin. Optom. **85** (2010). https://doi.org/10.2147/opto.s7021
17. Benjamin, W.J.: RGP material selection strategy and the Gel/RGP oxygen "Gap." Int. Contact Lens Clin. **20**, 200–203 (1993). https://doi.org/10.1016/0892-8967(93)90120-G
18. Doughman, D.: Is corneal refractive therapy a new modality whose time has come or repackaged orthokeratology whose time has passed? A view from an experienced corneal clinician. Eye Contact Lens **30**, 219–222 (2004). https://doi.org/10.1097/01.ICL.0000140227.51062.84
19. Hill, R.M.: Gas permeable perspecitves. J. BCLA. **8**, 27–29 (1985)
20. Gungor, B., Erdogan, H., Suner, S.S., Silan, C., Saraydin, S.U., Sahiner, N. Drug-impregnated contact lenses via supercritical carbon dioxide: A viable solution for the treatment of bacterial and fungal keratitis.Int. J. Pharm. **662**, 124505 (2024). https://doi.org/10.1016/j.ijpharm.2024.124505.3.x
21. Szczotka-Flynn, L.B., Patel, S.: Menicon Z rigid gas permeable lenses for keratoconus and irregular corneas: a retrospective case series. Eye Contact Lens **34**, 254–260 (2008). https://doi.org/10.1097/icl.0b013e31817f6db0
22. Port, M.J.A., Loveridge, R.: Menicon SP lenses for prolonged wear: a clinical report. J. Br. Contact Lens Assoc. **9**, 65–74 (1986). https://doi.org/10.1016/S0141-7037(86)80003-9
23. Akerman, A.: A theory on the role of wholesalers in international trade based on economies of scope. Can. J. Econ. **51**, 156–185 (2018). https://doi.org/10.1111/caje.12319
24. Key, J.E.: Development of contact lenses and their worldwide use. Eye Contact Lens **33**, 343–345 (2007). https://doi.org/10.1097/ICL.0b013e318157c230
25. Singh, J., Agrawal, K.K.: Polymeric materials for contact lenses. J. Macromol. Sci. Part C. **32**, 521–534 (1992). https://doi.org/10.1080/15321799208021431
26. FDA: Manufacturing/chemistry of contact lens materials. www.fda.gov (1987)
27. Shah, C., Sundar Raj, C.V., Foulks, G.N.: The evolution in therapeutic contact lenses. Ophthalmol. Clin. North Am. **16**, 95–101 (2003). https://doi.org/10.1016/S0896-1549(02)00066-4
28. González-Chomón, C., Concheiro, A., Alvarez-Lorenzo, C.: Soft contact lenses for controlled ocular delivery: 50 years in the making. Ther. Deliv. **4**, 1141–1161 (2013). https://doi.org/10.4155/tde.13.81
29. Michálek, J., Podešva, J., Dušková-Smrčková, M.: True story of Poly(2-Hydroxyethyl Methacrylate)-based contact lenses: how did it really happen. Substantia. **6**, 79–91 (2022). https://doi.org/10.36253/Substantia-1591
30. Wichterle, O., Lim, D.: Hydrophilic gels for biological use. Nature (1960)
31. Tran, N.P.D., Yang, M.C.: The ophthalmic performance of hydrogel contact lenses loaded with silicone nanoparticles. Polymers (Basel). **12** (2020). https://doi.org/10.3390/POLYM12051128
32. Abdulamier, A.A., Shaker, L.M., Al-Amiery, A.A.: Advancements in the chemistry of contact lenses: innovations and applications. Results Chem. **12**, 101872 (2024). https://doi.org/10.1016/j.rechem.2024.101872

33. Harbiyeli, I.I., Erdem, E., Isik, P., Yagmur, M., Ersoz, R.: Use of new-generation hybrid contact lenses for managing challenging corneas. Eur. J. Ophthalmol. **31**, 1802–1808 (2021). https://doi.org/10.1177/1120672120942757
34. Abdalla, Y.F., Elsahn, A.F., Hammersmith, K.M., Cohen, E.J.: Synergeyes lenses for keratoconus. Cornea **29**, 5–8 (2010). https://doi.org/10.1097/ICO.0b013e3181a9d090
35. Lipson, M.J., Musch, D.C.: Synergeyes versus soft toric lenses: vision-related quality of life. Optom. Vis. Sci. **84**, 593–597 (2007). https://doi.org/10.1097/OPX.0b013e318111ece4a
36. Morrison, C.: The Complete Guide to SynergEyes Hybrid Contact Lenses
37. Schiefer, U., Kraus, C., Baumbach, P., Ungewiß, J., Michels, R.: Refractive errors. Epidemiol. Effects Treatment Options, 693–702 (2016). https://doi.org/10.3238/arztebl.2016.0693
38. Swain, S., Sivakumar, R.: Prediction of myopia progression based on artificial intelligence model. Adv. Transdiscipl. Eng. **42**, 940–950 (2023). https://doi.org/10.3233/ATDE231034
39. Cuenca, B.: Top Contact Lenses for Astigmatism: A Comprehensive Review. https://opticaltraining.com/blog/top-contact-lenses-for-astigmatism-a-comprehensive-review/
40. Wolffsohn, J.S., Davies, L.N., Sheppard, A.L.: New insights in presbyopia: impact of correction strategies (2023)
41. Wolffsohn, J.S., Davies, L.N.: Presbyopia: effectiveness of correction strategies (2019)
42. Wojcik, K.A., Kaminska, A., Blasiak, J., Szaflik, J., Szaflik, J.P.: Oxidative stress in the pathogenesis of keratoconus and Fuchs endothelial corneal dystrophy. Int. J. Mol. Sci. **14**, 19294–19308 (2013). https://doi.org/10.3390/ijms140919294
43. Barrado-Navascúes, E., Vázquez-Fustes, M.J., Burgos-Martínez, M., Durán-Prieto, E., McGauley, J., Álvarez-Cerrato, M.: Vision evolution in the treatment of nystagmus with hydrophilic contact lenses (two clinical cases). Contact Lens Anterior Eye. (2018). https://doi.org/10.1016/j.clae.2018.03.092
44. Cox, I.G.: The why and wherefore of soft lens visual performance. Contact Lens Anterior Eye. **23**, 3–9 (2000). https://doi.org/10.1016/S1367-0484(00)80034-X
45. Fisher, K.: Clinical comparison of opaque tint soft contact lenses. Int. Contact Lens Clin. **23**, 46–49 (1996). https://doi.org/10.1016/0892-8967(96)00045-4
46. Baghban, R., Talebnejad, M.R., Meshksar, A., Heydari, M., Khalili, M.R.: Recent advancements in nanomaterial-laden contact lenses for diagnosis and treatment of glaucoma, review and update. J. Nanobiotechnol. **21**, 1–22 (2023). https://doi.org/10.1186/s12951-023-02166-w
47. Toffoletto, N., Saramago, B., Serro, A.P., Chauhan, A.: A physiology-based mathematical model to understand drug delivery from contact lenses to the back of the eye. Pharm. Res. **40**, 1939–1951 (2023). https://doi.org/10.1007/s11095-023-03560-7
48. Joshi, V.P., Singh, S., Thacker, M., Pati, F., Vemuganti, G.K., Basu, S., Singh, V.: Newer approaches to dry eye therapy: nanotechnology, regenerative medicine, and tissue engineering (2023)
49. Messmer, E.M.: Pathophysiology, diagnosis and treatment of dry eye. Dtsch. Arztebl. Int. (2015)
50. Satyam, K.: Therapeutic Contact Lenses: A Tool to Restore Eye Health. 1–7 (2023)
51. Mun, J., Mok, J.W., Jeong, S., Cho, S., Joo, C.K., Hahn, S.K.: Drug-eluting contact lens containing cyclosporine-loaded cholesterol-hyaluronate micelles for dry eye syndrome. RSC Adv. **9**, 16578–16585 (2019). https://doi.org/10.1039/c9ra02858g
52. Peng, C.C., Chauhan, A.: Extended cyclosporine delivery by silicone-hydrogel contact lenses. J. Control. Rel. (2011). https://doi.org/10.1016/j.jconrel.2011.06.028
53. Di Zazzo, A., Giannaccare, G., Villani, E., Barabino, S.: Uncommon Blepharitis. J. Clin. Med. **13**, 1–9 (2024). https://doi.org/10.3390/jcm13030710
54. Bernardes, T.F., Bonfioli, A.A.: Blepharitis. Semin. Ophthalmol. **25**, 79–83 (2010). https://doi.org/10.3109/08820538.2010.488562
55. Koay, S.Y., Larkin, D.F.P.: New pharmacological approaches for the treatment of neurotrophic keratitis. Front. Pharmacol. **13**, 796854 (2022). https://doi.org/10.3389/FPHAR.2022.796854/BIBTEX
56. Vieira, A.C., Mannis, M.J.: Ocular rosacea: common and commonly missed. J. Am. Acad. Dermatol. **69**, S36–S41 (2013). https://doi.org/10.1016/J.JAAD.2013.04.042

57. Taniguchi, E. V, Kalout, P., Pasquale, L.R., Kohane, D.S., Joseph, B.: Lenses for the treatment of glaucoma. **5**, 1077–1083 (2021). https://doi.org/10.4155/tde.14.76.Clinicians
58. Lim, L., Lim, E.W.L.: Therapeutic contact lenses in the treatment of corneal and ocular surface diseases—a review. Asia-Pacific J. Ophthalmol. **9**, 524–532 (2020). https://doi.org/10.1097/APO.0000000000000331
59. Zhao, L., Song, J., Du, Y., Ren, C., Guo, B., Bi, H.: Therapeutic applications of contact lens-based drug delivery systems in ophthalmic diseases. Drug Deliv. **30** (2023). https://doi.org/10.1080/10717544.2023.2219419
60. Huang, M.: Contact Lenses 101, Part 2: Contact Lens Designs
61. Backhouse, S., Fox, S., Ibrahim, B., Phillips, J.R.: Peripheral refraction in myopia corrected with spectacles versus contact lenses. Ophthalmic Physiol. Opt. **32**, 294–303 (2012). https://doi.org/10.1111/j.1475-1313.2012.00912.x
62. Moore, K.E., Benoit, J.S., Berntsen, D.A.: Spherical soft contact lens designs and peripheral defocus in myopic eyes. In: Optometry and Vision Science (2017)
63. González-Méijome, J.M., Faria-Ribeiro, M.A., Lopes-Ferreira, D.P., Fernandes, P., Carracedo, G., Queiros, A.: Changes in peripheral refractive profile after orthokeratology for different degrees of myopia. Curr. Eye Res (2016). https://doi.org/10.3109/02713683.2015.1009634
64. Zhang, G.Y., Ye, L., Wang, W.J., Guo, Y.M., Wei, J.H., Ren, M.X.: Effectiveness of toric soft contact lenses for vision correction in patients with different degrees of astigmatism: a real-world study. Int. J. Ophthalmol. **16**, 1845–1853 (2023). https://doi.org/10.18240/IJO.2023.11.17
65. Filho, R.G., Giovedi, M.A., Nichols, J.J.: Design and nomenclature of contact lenses. In: Contact Lenses Ophthalmic Practice, pp. 1–6 (2004). https://doi.org/10.1007/0-387-21758-4_1
66. Herranz, R., Rodrguez, G., de Juan Herrez, V.: Contact lens correction of regular and irregular astigmatism. Astigmatism—Opt. Physiol. Manag (2012). https://doi.org/10.5772/19001
67. Gustafson, B., Whaley, S., Giedd, B., Montaquila, S., Edwards, B., Subbaraman, L.N.: Short-term fit assessment of a novel daily disposable, toric silicone hydrogel contact lens. Clin. Ophthalmol. **16**, 3721–3730 (2022). https://doi.org/10.2147/OPTH.S370151
68. Schornack, M.: Prescription and management of contact lenses in patients with monocular visual impairment. Optometry **78**, 652–656 (2007). https://doi.org/10.1016/j.optm.2007.02.022
69. Charman, W.N.: Developments in the correction of presbyopia I: Spectacle and contact lenses. Ophthalmic Physiol. Opt. **34**, 8–29 (2014). https://doi.org/10.1111/opo.12091
70. Remón, L., Pérez-Merino, P., MacEdo-De-Araújo, R.J., Amorim-De-Sousa, A.I., González-Meíjome, J.M.: Bifocal and multifocal contact lenses for presbyopia and myopia control. J. Ophthalmol. **2020** (2020). https://doi.org/10.1155/2020/8067657
71. Kim, D.Y., Ahn, H., Lee, S., Jun, I., Seo, K.Y., Kim, S.Y., Kim, T.I.: Efficacy of a new refractive multifocal contact lens for presbyopia. Sci. Rep. **14**, 1–9 (2024). https://doi.org/10.1038/s41598-024-55918-5
72. Kollbaum, P.S., Bradley, A., Thibos, L.N.: Comparing the optical properties of soft contact lenses on and off the eye. Optom. Vis. Sci (2013). https://doi.org/10.1097/01.opx.0000434275.93435.da
73. Zheleznyak, L., Jung, H., Yoon, G.: Impact of pupil transmission apodization on presbyopic through-focus visual performance with spherical aberration. Investig. Ophthalmol. Vis. Sci (2014). https://doi.org/10.1167/iovs.13-13107
74. Fedtke, C., Bakaraju, R.C., Ehrmann, K., Chung, J., Thomas, V., Holden, B.A.: Visual performance of single vision and multifocal contact lenses in non-presbyopic myopic eyes. Contact Lens Anterior Eye (2016). https://doi.org/10.1016/j.clae.2015.07.005

Chapter 4
The Criteria and Properties of Contact Lenses

4.1 Industry Standards and Quality Criteria for Contact Lenses

The industry standards for determining the criteria for contact lenses are primarily governed by the International Organization of Standardization (ISO)—ISO 18369, which covers rigid (hard) corneal and scleral contact lenses, as well as soft contact lenses. This standard outlines and defines the terms related to the physical, chemical, and optical properties of contact lenses. ISO 18369 is divided into four parts: ISO 18369-1, ISO 18369-2, ISO 18369-3, and ISO 18369-4. The first part (ISO 18369-1) focuses on the vocabulary and classification systems used in the contact lens field, along with recommendations for labelling specifications. Meanwhile, ISO 18369-2 covers tolerances for contact lens properties, ISO 18369-3 provides measurement methods, and ISO 18369-4 details the physicochemical properties of contact lens materials.

The terminology and definitions widely used in the contact lens industry are outlined in the first part of ISO 18369 (ISO 18369-1), which also includes international symbols and abbreviations. This standard comprehensively covers the essential terms, including those related to basic concepts, contact lens parameters and design, scleral lenses and shells, and definitions of material properties. Furthermore, the classification of contact lens materials is detailed in this standard, providing the basis for naming materials such as lotrafilcon A, lotrafilcon B, and galyfilcon A. Each of these names follows a systematic six-part coding structure, which categorizes contact lens materials based on several key characteristics. This coding structure includes the following components: (prefix)(stem)(series)(group suffix)(Dk range)(modification code), laying the foundation for material identification and classification.

The naming of these contact lens materials truly depends on the contact lens properties, whether it is from hydrogel or non-hydrogel material, ionic or non-ionic material, the water content and the oxygen permeability of the material (refer the

© The Author(s), under exclusive license to Springer Nature Singapore Pte Ltd. 2025
N. Adrus et al., *Contact Lenses*, Series in BioEngineering,
https://doi.org/10.1007/978-981-96-3636-5_4

classification of soft contact lenses according to US FDA in Sect. 3.1.2). In the six-part coding system, the prefix designates the material's chemical formulation and is administered by the United States Adopted Names (USAN) Council. While the use of prefix is mandatory in the United States, it is optional in the other region. Some common prefixes include *lotra*, *galy*, and *com*. The next part of the coding system is the stem, which falls into one of two categories: *filcon* for materials with 10% or more water by mass and *focon* for those with less than 10% water by mass. This structured naming convention allows for precise identification of contact lens materials based on their formulation and properties.

Next, in the six-part coding system, the USAN Council assigns a series suffix, particularly when altering the monomer ratio in an existing contact lens material to create a new material. For example, the capital letter "A" is added after the stem designation for the initial material. When the monomer ratio is adjusted again, the next letter in the alphabet is used. Hence, these series suffixes distinguish between contact lens materials with identical monomer components but different ratios. The group suffix in this coding system represents the material's water content and ionic properties for *filcon* materials according to ISO 18369-4, and is using Roman numerals. While in *focon* materials, the group suffix indicates the presence or absence of silicone/fluorine. The oxygen permeability (Dk) is another crucial parameter, and its range is represented by a numeric designation within the coding structure, categorizing the material's ability to allow oxygen to pass through, which is vital for contact lens performance. The last suffix in the six-part coding is the modification code, indicated with a lowercase letter "m", used only if the contact lens has undergone surface modifications. The surface modifications may include processes like plasma treatment, acid/base hydrolysis, or incorporation of a material that migrates to the lens surface, altering the surface characteristics compared to the bulk material.

The second part of ISO 18369 (ISO 18369-2) provides detailed specifications regarding the tolerance limit for the optical and physical parameters of rigid corneal, rigid scleral, and soft contact lenses at the time of manufacture. It sets the acceptable dimensional and optical property ranges, as well as material specifications for various types of contact lenses. These parameters are essential for ensuring consistency and performance in lenses. ISO 18369-2 also links to the tests described in ISO 18369-3 and ISO 18369-4, which provide further guidance on measuring these properties. The conformance criteria for testing contact lens are included in this document, which are divided into three methods: 100% inspection, acceptance sampling, and statistical estimation of performance.

Part 3 of ISO 18369-3 delves into the standardized methods for measuring the physical and optical properties of contact lenses (ISO 18369-3). This standard also provides comprehensive specifications for the instruments employed in these assessments, as summarized in Table 4.1. Specifically, this part details the measurement procedures for critical parameters such as radius of curvature, back vertex power, diameter and thickness, along with guidelines for inspecting of edges, inclusions, surface imperfections. These critical parameters, specifically the dimensions of the contact lens are further discussed in Sect. 4.2.1. Furthermore, it encompasses methods for determining spectral and luminous transmittances. Each measurement protocol is accompanied by guidance on suitable instrumentation, encompassing principles of operation, specifications, and calibration methods. This meticulous standardization, as presented in ISO 18369-3, serves as an invaluable resource for researchers, enabling informed decisions regarding the acquisition on appropriate instrumentation. Moreover, Table 4.2 provides a consolidated overview of the instruments and test methods employed in ISO 18369-3 for various contact lens properties, categorized by lens type and reproducibility [1].

For assessing physicochemical properties, Part 4 of ISO 18369 (ISO 18369-4) describes the methods for evaluating key characteristics of contact lens including extractables (compounds that may leach from the lens material), lens flexural deformation and rupture, oxygen permeability, refractive index and water content [2]. This part comprehensively outlines the principles, apparatus, reagents used, sample conditions, test procedures, and calculation of results for each contact lens property. The instruments and test methods used to assess different contact lens properties in ISO 18369 Part 4, summarized in Table 4.3.

4.2 Essential Characteristics of Contact Lenses

The essential characteristics of contact lenses are categorized into physical, optical, and physicochemical properties. The physical properties of contact lens metrology include the contact lens parameters and design, radius of curvature, diameter, and thickness, which are detailed in Sect. 4.2.1 under lens metrology. Physicochemical properties, which cover spectral and luminous transmittance, oxygen permeability, refractive index, equilibrium water content, elastic modulus, and extractable profiles, are discussed in Sect. 4.2.2 as key properties of contact lenses.

4.2.1 Insights in Lens Metrology

Generally, contact lens metrology consists of the main and specific components, as illustrated in Fig. 4.1, including the contact lens axis, base curve, optical curve, rolled edge, blend (the transition point between radii of curvature), and peripheral curve [3]. Among these, the rolled edges, shown in Fig. 4.1a, play a crucial role in enhancing comfort by ensuring a smooth transition between the lens and the eye. The geometry of recent contact lenses can be further divided into two, which are bi-curve and multi-curve lenses. In Fig. 4.1b, a bi-curve lens features two distinct curves: a primary base curve and a flatter peripheral curve. Furthermore, Fig. 4.1b highlights the two key diameter measurements of a contact lens: the total diameter, spanning from one edge of the rolled edge to the other, and the optic zone diameter, representing the distance between the junctions of the lens. In contrast, a multi-curve contact lens, as shown in Fig. 4.1c, has a base curve with three or more peripheral curves. Specifically, a tri-curve contact lens has two peripheral curve radii: the peripheral curve radius and

Table 4.1 Specifications of instruments for the measurement of contact lens properties [1]

Properties and instruments	Specification
1. Radius of curvature	
(a) Optical microspherometer	• Fitted with a vertical illuminator and a target with a fine focus adjustment • Objective lens magnification: × 6.5 with a numerical aperture of not less than 0.25 • Total magnification: × 65. The real image of the target formed by the microscope shall not be greater than 1.2 mm in diameter • Gauge interval scale: not more than 0.02 mm • Gauge accuracy: ± 0.010 mm for readings for 2.00 mm or more at a temperature of 20 °C ± 5 °C • Gauge repeatability: ± 0.003 mm • Attachment: contact lens holder that is capable of holding the contact lens surface in a reference plane that is normal to the optic axis of the instrument
(b) Ophthalmometer with contact lens attachment	• Ophthalmometer with a lighted target positioned to ensure a reflection from an optical surface placed perpendicular to the axis of the optical system • A special lens-holding attachment is necessary so that the contact lens is held in the proper location and orientation • Capable of assessing the size of the reflected image of a target of fixed size and distance, or target size must be sufficiently adjustable with a fixed doubling system so as to attain a reflected image of fixed size • The ophthalmometer shall be capable of measurement of the two primary meridians of a toric surface • Total magnification: × 20 or higher • Scale interval: 0.02 mm or lesser. (Diopters = 0.25 D)
2. Sagittal depth	
(a) Optical comparator	• Minimum magnification of × 10 Incorporated with a wet cell with a hollow cylindrical contact lens support • A graticule marked in minimal increments of 0.01 mm shall be used so that sagittal depth can be visually determined with repeatability of ± 0.02 mm or better when used to measure sagittal depths within the central portion of the contact lens (when the contact lens is properly centered on its support). The instrument shall have a means of controlling the wet cell temperature
(b) Spherometer	• The spherometer projects the profiles of the contact lens, lens support and probe onto a screen • Magnification specification: × 10 or higher • Analogue or digital gauges precision: 0.01 mm or higher

(continued)

Table 4.1 (continued)

Properties and instruments	Specification
(c) Ultrasonic depth gauge	• Sagittal depth measurement assisted by ultrasound unit • The ultrasonic transducer shall be fitted under the center of the contact lens support • Frequency: greater than 18 MHz Bandwidth of 1.4 f ± 5 MHz Beamwidth: 2.0 mm or less at focus Focal length: 25 mm to 50 mm • The interval timer: able to record times to 0.01 μs • The ancillary apparatus is able to reject or ignore the unwanted reflection signal • Repeatability of the ultrasound unit with wet cell: ± 0.02 mm or better when used to measure sagittal depths within the central portion of the contact lens (when the contact lens is properly centred on its support)
3. Back vertex power	
(a) Focimeter	• Minimum range of − 20.00 D to + 20.00 D • Minimum measuring accuracy of ± 0.06 D, capable of manual focusing
4. Total diameter	
(a) V-groove gauge method	• Capable of measuring to ± 0.05 mm over a range from 7 to 11 mm • The groove is engraved at its edge or center, with a scale for reading off diameters • The scale should be inscribed for diameters from 7.0 to 11.0 mm • The graduation lines should be placed at diameter intervals of 0.10 mm with a longer line at each interval of 0.50 mm and a more prominent line at each interval of 1.00 mm
(b) Projection comparator method	• Capable of measuring to ± 0.05 mm over a range of 0–7 mm • The scale of the screen represents a linear magnification of at least × 15 and permits measurement accuracy of 0.05 mm for the contact lens diameter • The apparatus has a telecentric path of rays, which is ensured by positioning the diaphragm in the rear focal plane of the objective • The diameter of the lens is either taken from linear markings on the screen or from a marked glass scale (similar to a microscope graticule) under the lens within the support, which is projected onto the screen

(continued)

Table 4.1 (continued)

Properties and instruments	Specification
5. Thickness	
(a) Dial gauge and "low force" dial gauge	Dial gauge • The measuring face of the dial gauge: spherical with a radius between 1.2 and 5.0 mm • For the dial gauge, apply force not exceeding 1.4 N on the lens • For low force dial gauge, apply force: not exceed 0.015 N (1.5 g) when displaced by 1 mm from its unrestrained position (i.e. rest) • Minimum scale increments: not more than 0.01 mm • Reproducibility: equal to or better than 0.01 mm over the range 0–5 mm • Capable of being set to zero using a fixed stop so that the contact lens surfaces at the point where the thickness is to be measured shall be in contact with the adjustable stylus and the fixed stop

the intermediate curve radius. The intermediate curve is located between the base and the peripheral curves. For a standard tri-curve contact lens, the peripheral curve radius is typically 12.25 mm, and the intermediate curve is 1 mm flatter than the base curve. Additionally, the space between the rolled edge and the junction constitutes the peripheral curve width of the lens.

Besides, the center thickness of a contact lens is another important component, where it separates the anterior and posterior surfaces at the geometric center of a lens, depicted in Fig. 4.1a. Interestingly, the lens power is determined by the center thickness, where thinner center results in higher minus power, and vice versa. Furthermore, the optic curve of a contact lens illustrated in Fig. 4.1a is where the optic zone is located with its refractive power. Meanwhile, the sagittal depth of the contact lens is the distance from a flat plane to the back surface of the center of the lens.

The sagittal depth of a contact lens, a key determinant of its overall geometry, is influenced by its diameter and curvature. A lens with the same diameter but different sagittal depth (longer radius) produces a greater "vaulting" of the lens, commonly called the sagittal vault. Figure 4.2 depicts a lens with a constant diameter but a longer radius (e.g., 7.8–8.4 mm), resulting in a shortened sagittal vault and a flatter lens profile. Conversely, a lens with a constant diameter but a shorter radius leads to an increased sagittal depth. Alternatively, Fig. 4.3 portrays a lens with a constant radius of curvature but an increased diameter (e.g., 13–15 mm), demonstrating a corresponding increase in sagittal height and a steeper lens profile.

Table 4.2 Instruments and test methods used to measure the dimension and properties of contact lenses as described in Part 3 of ISO 18369

Contact lens dimension*	Instrument/Test method	Type of contact lens	Reproducibility, R/ ISO part 2
Radius of curvature	Optical microspherometer	Spherical rigid lenses	± 0.015 mm in air
	Ophthalmometer with contact lens attachment	Spherical rigid lenses	± 0.015 mm in air
		Spherical rigid lenses	± 0.025 mm in saline solution
		Spherical hydrogel lenses (38% water content, $t_c > 0.1$ mm)	± 0.050 mm in saline solution
	Sagittal height method	Hydrogel lenses (38% water content, $t_c > 0.1$ mm)	± 0.050 mm in saline solution
		Hydrogel lenses (55% water content, $t_c > 0.1$ mm)	± 0.100 mm in saline solution
		Hydrogel lenses (70% water content, $t_c > 0.1$ mm)	± 0.200 mm in saline solution
Sagittal depth	Optical comparator Spherometer Ultrasound unit	Hydrogel lenses (< 38% water content)	± 0.050 mm
		Hydrogel lenses (38–54% water content)	± 0.100 mm
		Hydrogel lenses (> 54% water content)	± 0.200 mm
Back vertex power	Focimeter (determined in air)	Rigid, hydrogel (including spherical and toric), and non-hydrogel lenses	± 0.400 D in air
	Moiré deflectometer (determined in saline solution)	Spherical and toric hydrogel contact lenses	± 0.2528 in saline solution
	Hartmann method (determined in saline solution)	Spherical and toric hydrogel contact lenses	± 0.0708 in saline solution
Total diameter	V-groove gauge method	Rigid contact lenses	± 0.05 mm
	Projection comparator method	Soft contact l lenses	± 0.05 mm
Zone diameters (optic zone, secondary and peripheral curve) and width	Hand-held scale magnifier	Rigid contact lenses	N/A

(continued)

Table 4.2 (continued)

Contact lens dimension*	Instrument/Test method	Type of contact lens	Reproducibility, R/ ISO part 2
	Projection comparator method	Soft contact l lenses	N/A
Thickness	Dial gauge	Rigid contact lenses	± 0.010 mm
	"Low force" dial gauge	Soft contact l lenses	± 0.017 mm
Edges inspection	visual examination under direct illumination (× 7 magnification)	Rigid and soft contact lenses	N/A
Inclusions and surface imperfections	Enlarging magnifiers	Rigid and soft contact lenses	N/A
Spectral and luminous transmittance	UV–visible spectroscopy	Rigid and soft contact lenses	N/A

*Description and illustration of contact lens dimensions are provided in Sects. 4.2.1 and 4.2.2

Table 4.3 Instruments and test methods used to measure the chemical properties of contact lenses as described in Part 4 of ISO 18369

Contact lens properties	Instrument/Test method	Application
Extractables	Soxhlet extraction	Rigid, hydrogel, and non-hydrogel materials
Lens flexural deformation and rupture	Apparatus which allows load and flexural deformation	Rigid contact lenses
Oxygen permeability	Polarographic method	All type of materials with oxygen permeability from 0 to 145
	Coulometric method	Non-hydrogel materials
Refractive index	Calibrated Abbe Refractometer or equivalent refractometer	Rigid, hydrogel, and non-hydrogel materials
Water content	Gravimetric method	Rigid, hydrogel, and non-hydrogel materials
	Refractive index method	Rigid, hydrogel, and non-hydrogel materials

4.2.2 Key Properties of Contact Lenses

Contact lenses must possess specific properties within accepted ranges to ensure optimal performance and safety. These critical properties include spectral and luminous transmittance, oxygen permeability, refractive index, equilibrium water content, modulus, and coefficient of friction. Table 4.4 provides a summarized overview of the acceptable ranges for each of these essential contact lens properties.

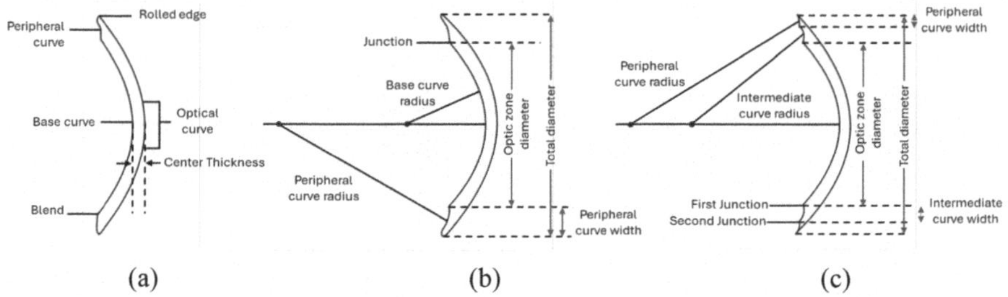

Fig. 4.1 **a** Main parts of contact lens: base curve, optical curve, rolled edge, blend, peripheral curve and center thickness, **b** depiction of additional design elements, including total diameter, optic zone diameter, and peripheral curve width, and **c** illustration of tri-curve lens design elements

Spectral transmittance and luminous transmittance are the terms used to measure the transmittance of contact lenses. By definition, spectral transmittance is the ratio of transmitted spectral radiant flux to the incident spectral radial flux. The spectral transmittance property is significant for visual performance [12] where it measures the UVA, UVB, UVC, and visible transmittances of the lens. On the other hand, luminous transmittance is significant for contact lens users to detect objects with low contrast. To obtain the luminous transmittance, the spectral transmittance at wavelength spectrum of 280–780 nm were measured and integrated in small increments (< 10 nm) and the radiant flux was weighted by the relative spectral luminous efficiency function of the eye. The measurements and transmittance are specified in saline conditions.

Another crucial property of contact lenses is oxygen permeability, which refers to how easily oxygen can pass through the lens material. It is quantified by the permeability coefficient (Dk), which is the product of diffusion (D) and solubility of oxygen in the lens material (k). The standard for measuring Dk is Barrer, where 1 Barrer = [cm^3 (STP oxygen)cm^2/cm^3 (material s mm Hg] $\times$ 10^{-11} [13]. While Dk measures the inherent ability of the material to allow oxygen flow, the clinically relevant factor is oxygen transmissibility (Dk/t), which take into account for both the material's permeability and the lens thickness (t). A thinner lens with the same Dk will transmit more oxygen to the cornea. Therefore, Dk/t provides a more practical assessment of the oxygen that reaches the eye while wearing lens [11]. According to recommendations, a Dk/t value of $24.1 \pm 2.7 \times 10^{-9}$ units is advised for soft contact lenses to minimize the corneal swelling and discomfort [14].

The amount of oxygen that reaches the cornea is vital for preserving its physiological integrity. The cornea relies on oxygen absorbed from tears and the atmosphere because it lacks its own blood supply. When contact lenses are worn, they create a barrier that can reduce the amount of oxygen reaching the cornea, potentially leading to discomfort, redness, poor vision, and in severe cases, hypoxic stress. Maintaining sufficient oxygen levels is therefore critical, not only for comfort but also for the overall health of the eye. As such, oxygen permeability is a key property

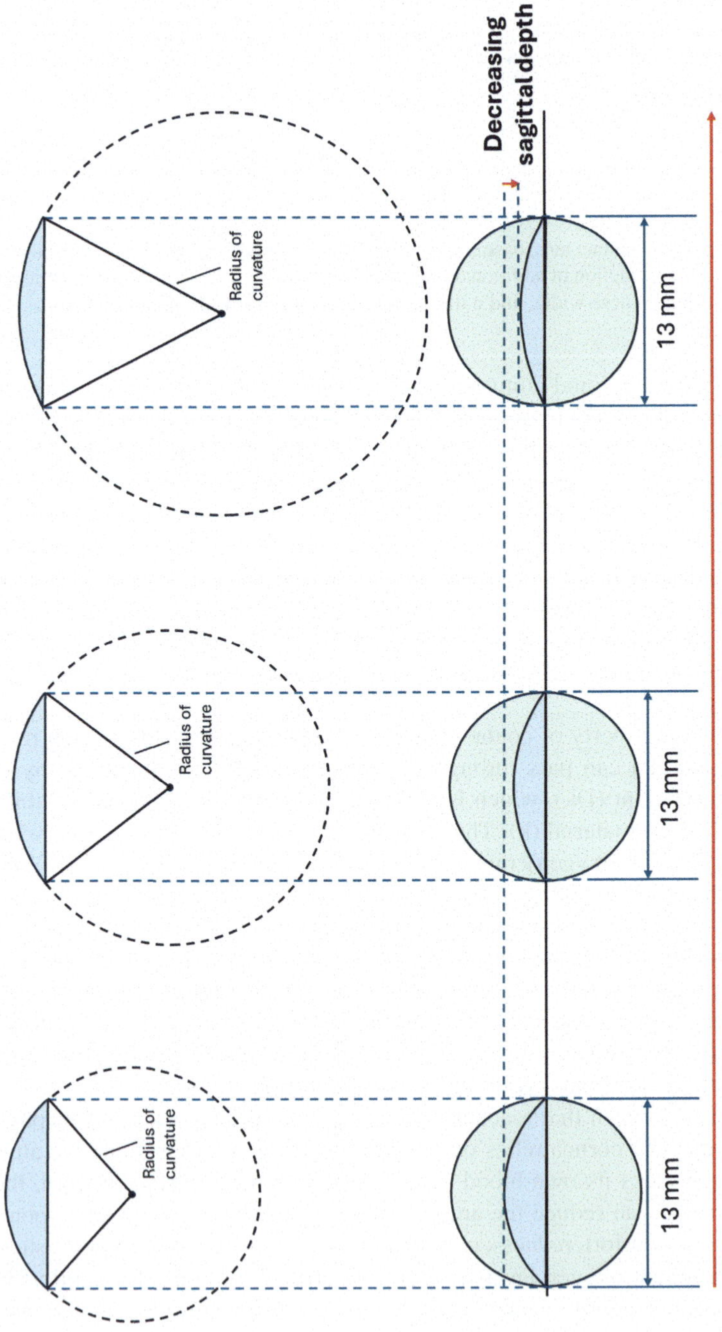

Fig. 4.2 The effect of increasing the radius of curvature at a constant diameter of 13 mm, resulting in a flatter contact lens

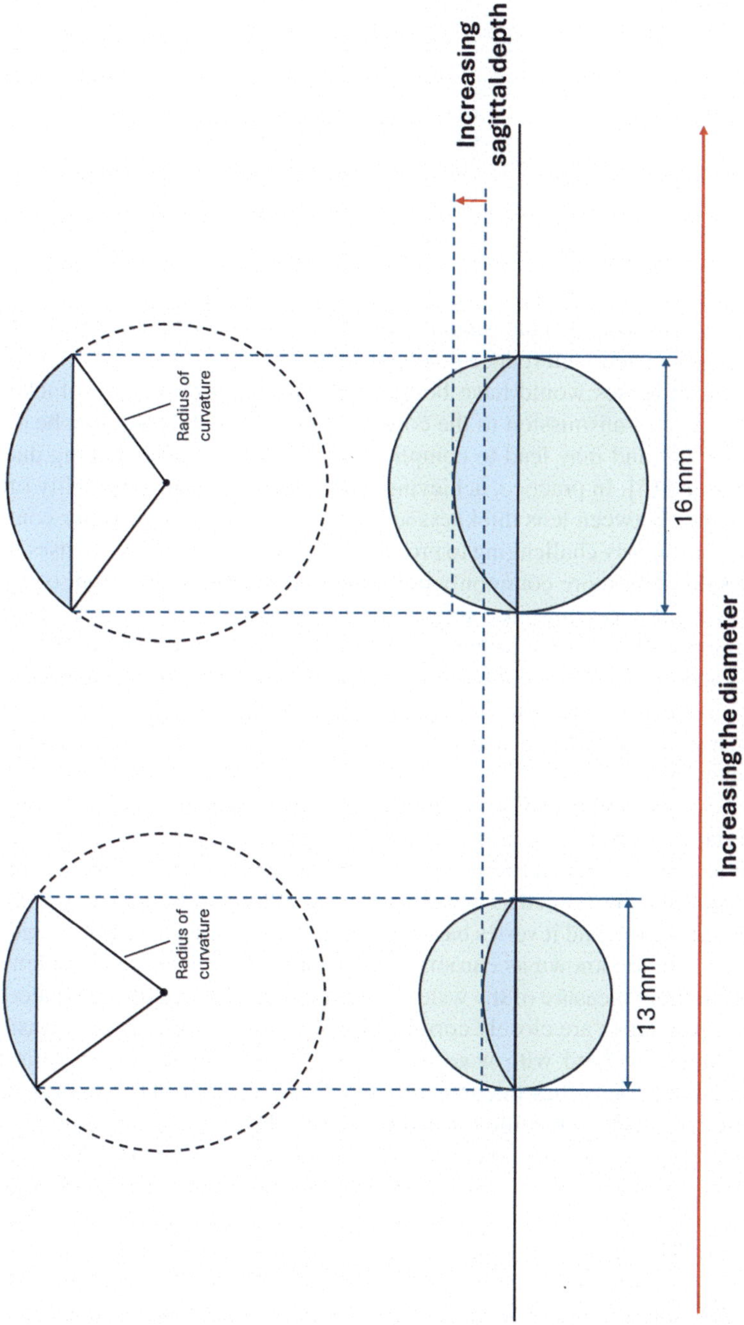

Fig. 4.3 The effect of increasing the diameter of contact lens at a constant radius of curvature, resulting in a greater sagittal depth

Table 4.4 Acceptable ranges for the physical properties of contact lenses

Physical properties	Range	References
Transparency	> 90% transmittance	[4]
Refractive index	1.40 to 1.43	[5, 6]
Oxygen permeability	Open eyes: 25 Dk units Closed eyes: 75–200 Dk units	[7, 8]
Tensile strength (Young's modulus)	Hydrogel: 0.2–0.8 MPa Silicone hydrogel: 0.4–1.1 MPa	[9, 10]
Lens diameter	13.8–14.5 mm	[11]

of contact lenses, ensuring that the wearer's cornea remains healthy while preventing complications associated with reduced oxygen supply.

Ideally, contact lenses would have both a high Dk and a thin center thickness to maximize oxygen transmission to the cornea. However, such lenses can be difficult to manufacture and may lead to complications such as corneal staining due to rapid dehydration [15]. In practice, achieving optimal oxygen transmissibility often requires a balance between lens thickness and Dk, as very thin, high-water-content lenses can be particularly challenging to produce [16]. As a result, thicker lenses with limited Dk/t values are more commonly used in clinical settings. Different types of contact lenses exhibit varying Dk values, with the silicon hydrogel lenses having the highest range (60–140 Dk), followed by the rigid gas permeable (RGP) lenses (18–163 Dk) and the conventional hydrogel lenses (9–92 Dk). For further insights, Tables 4.5, 4.6 and 4.7 summarized the comparison of the commercially available RGP, hybrid, traditional hydrogel and silicone hydrogel contact lenses based on their material properties.

Besides, refractive index (RI) and equilibrium water content (EWC) are fundamental material properties of contact lenses that significantly impact both optical performance and physiological response, reflecting the underlying polymer composition [5]. The term RI is defined as the ratio of speed of light in vacuum to the speed of light in the substance, and it varies based temperature and wavelength. Changes in RI with wavelength are known as chromatic dispersion [33]. In soft contact lenses, RI is often an indirect measure of the water content, especially in hydrogel materials where the two properties are closely correlated. EWC, on the other hand, measures a hydrogel's ability to bind with water, which is essential in achieving desirable lubricity and clinical outcomes in contact lenses. Water content plays an important role in lens functionality, impacting ion and gas permeability, mechanical properties surface characteristics and biocompatibility. Maintaining consistent EWC during wear is critical for optimizing comfort and ensuring long-term success of contact lenses.

In addition to RI and EWC, the mechanical properties of contact lenses are essential for ensuring comfort and effective performance. Factors such as eye movement, repeated application and removal can exert stress on contact lens materials, influencing clinical performance by affecting on-eye movement, wettability and fit. These

Table 4.5 The properties of RGP and hybrid lenses from several manufacturers in terms of oxygen permeability, EWC, elastic modulus, and contact angle

Brand	Manufacturer	Material (USAN)	Dk (Barrer)	EWC (%)	Elastic modulus (MPa)	Contact angle (°)	References
Boston ES®	Bausch & Lomb	Enflufocon A	18	–	–	52	[17]
Boston EO®	Bausch & Lomb	Enflufocon B	58	–	–	49	[17]
Boston XO®	Bausch & Lomb	Hexafocon A	100	–	–	49	[17]
Boston XO_2®	Bausch & Lomb	Hexafocon B	141	–	–	38	[17]
Menicon Z®	Menicon	Tisilfocon A	163	–	–	24	[17]
Duette	Synergeyes	RGP center: Petrafocon A Soft skirt: Hem-larafilcon A	RGP center: 130 Soft skirt: 84	RGP center: – Soft skirt: 32	–	–	[18]
UltraHealth	Synergeyes	RGP center: Si-150 Soft skirt: SI–H	RGP center: 130 Soft skirt: 84	RGP center: – Soft skirt: 32	RGP center: 1314 Soft skirt: 0.8	RGP center: 34° Soft skirt: 35°	[19]
ClearKone	Synergeyes	RGP center: Paflufocon D Soft skirt: Hem-iberfilcon A	RGP center: 100 Soft skirt: 9.3	RGP center: 1 Soft skirt: 27	RGP center: 1150 Soft skirt: 0.8	RGP center: 42° Soft skirt: –	[19]

stresses may also contribute to discomfort, or in severe cases, ocular lesions [34–36]. The modulus, a key parameter for describing a lens's mechanical properties, represents the material's elasticity [9]. Modulus is typically expressed in Mega Pascal (MPa), which is convertible between units in the form of force per unit area (1 MPa $= 10^6$ Nm^{-2} = 145.04 psi = 107 dynes cm^{-2}). Studies suggest that a modulus range of 0.4 to 1.1 MPa (Table 4.4) contributes to greater comfort by reducing frictional forces during blinking, minimizing irritation from the interaction between the lens and the eyelid.

Beyond elasticity, maintaining low frictional losses during the blinking cycle are also essential for a comfortable wearing, as the eyelid moves back and forth [37]. The presence of a contact lens between the cornea and the eyelid, however disrupts the natural function of the tear film [37]. This disruption alters the tear-exchange rate

Table 4.6 The properties of conventional hydrogel contact lenses including the oxygen permeability, center thickness, EWC, and its ionicity

Brand	Manufacturer	Material (USAN)	Dk	Centre thickness @− 3.00 D (mm)	EWC (%)	Ionic	References
Alden Flexlens	Bausch & Lomb	Hioxifilcon B	15	0.1	49	No	[20]
Softlens 38 Softlens multifocal	Bausch & Lomb	Polymacon	9	0.035	38	No	[21]
Purevision® 2	Bausch & Lomb	Balafilcon A	91	0.07	36	Yes	[22]
Soflenstoric	Bausch & Lomb	Alphafilcon A	32	0.19	66	No	[23]
Softlens Daily Disposable	Bausch & Lomb	Hilafilcon B	22	–	59	No	[24]
Biotrue® Oneday	Bausch & Lomb	Nesofilcon A	42	–	78	No	[25]
Focus Dailies	Alcon	Nelfilcon A	26	0.10	69	No	[26]
Proclear 1-Day Misight 1-Day	CooperVision	Omafilcon A	33	0.09	60	No	[27]
Proclear Multifocal, Sphere and Toric	CooperVision	Omafilcon B	34	–	62	No	[28]
Biomedics	CooperVision	Ocufilcon D	19.7	–	55	Yes	[29]
1-Day Acuvue Moist Acuvue 2	Johnson & Johnson	Etafilcon A	21	0.08	58	Yes	[30]
Kontur	Kontur	Methafilcon A	18	–	55	Yes	[31]
Freshlook Colourblends Dimensions	Aqualens	Phemfilcon A	16	–	55	Yes	[31]

and affects the stability and activity of both lipids and proteins in the tear film as they interact with the lens surface [38, 39]. In addition, the coefficient of friction (CoF) is defined as the ratio of frictional to normal force between the two surfaces, is used to assess the tribological behavior of a material. Most of the tribological studies for silicone hydrogel contact lenses have been based on the reported values of contact pressures ranging between 3 and 5 kPa, with the sliding speed around 12 cm/s [40].

The key characteristics of contact lenses, such as spectral and luminous transmittance, oxygen permeability, water content, refractive index, modulus, and coefficient of friction, are crucial for assessing their effectiveness. These properties are of particular interest to manufacturers looking to improve their products and researchers

Table 4.7 The properties of commercially available SiHy contact lenses (adapted with permission from González-Méijome et al. [32])

Brand	Manufacturer	Material (USAN)	Generation	Dk	Dk/t (barrer/cm)	Centre thickness @ – 3.00 D (mm)	EWC (%)	Ionic (FDA Group)	Elastic modulus (MPa)	Friction coefficient	Contact angle (°)
Air Optix Night & Day	Alcon Vision Care	lotrafilcon A	First	140	175	0.08	24	No (I)	1.4	0.394	80
Air Optix	Alcon Vision Care	lotrafilcon B	Second	110	138	0.08	33	No (I)	1.2	0.361	78
Dailies Total 1	Alcon Vision Care	delefilcon A	First	140	156	0.07	32	No (II)	0.7	–	–
Purevision	Bausch & Lomb	balafilcon A	First	99	110	0.09	36	Yes (III)	1.1	0.443	95
Acuvue Advance	J&J Vision Care	galyfilcon A	Second	60	86	0.07	47	No (I)	0.43	0.028	65
Acuvue OASYS	J&J Vision Care	senofilcon A	Second	103	147	0.07	38	No (I)	0.72	0.024	68
Acuvue TruEye	J&J Vision Care	narafilcon A	Third	100	125	0.085	46	No (I)	0.66	–	–
Biofinity	Cooper–vision	comfilcon A	Third	128	160	0.08	48	No (I)	0.8	0.057	–
Avaira	Cooper–vision	enfilcon A	Third	100	125	0.08	46	No (I)	0.5	0.018	–
PremiO	Menicon	asmofilcon A	Third	129	161	0.08	40	No (I)	0.91	0.207	–
Definitive	Contamac	efrofilcon A	Third	60	75	0.08	74	No (II)	0.35	–	–
Total30®	Alcon Vision Care	lehfilcon A	Fourth	–	154	0.08	55	No (II)	0.60	–	

seeking to develop innovative materials and technologies for contact lenses. For further insights, Sect. 4.3 provides a comprehensive comparison of commercially available contact lenses, followed by an exploration of cutting-edge research in lens properties in Sect. 4.4.

4.3 Comparison of Contact Lenses Characteristics in the Market

The contact lens market presents a diverse selection of products, with various brands offering lenses with unique compositions and performance characteristics. While RGP lenses are still available from manufacturers like Bausch & Lomb and Menicon (as summarized in Table 4.5), most modern lenses utilize silicon monomers in diverse formulations. As discussed in previous sections, key factors like modulus, coefficient of friction, oxygen permeability, wettability, refractive index, and equilibrium water content play crucial roles in determining lens performance and wearer comfort. This section presents a comparative analysis of these properties in commercially available contact lenses, providing insights into how they align with established benchmarks. By exploring these characteristics, we aim to provide a comprehensive understanding of the current contact lens landscape, ultimately facilitating both practitioners and users in making informed decisions to choose the most suitable lenses for their specific needs.

RGP lenses are renowned for their exceptional oxygen permeability and higher transparency compared to other lens types. In the following Table 4.5, the properties of various RGP lenses from Bausch & Lomb and Menicon, including the commercial names Boston ES, EO, XO, XO2, and Menicon Z are summarized. The oxygen permeability for these RGP lenses ranges from 18 to 163 Dk, providing superior breathability. Additionally, RGP lenses exhibit higher hydrophilicity, with wetting angles ranging from 38° to 52°. In the case of hybrid contact lenses, a RGP optical zone center exhibits excellent oxygen permeability, generally between 100 and 130 Dk, while the silicone hydrogel skirt has an oxygen permeability ranging from 9.3 to 84 Dk [41]. Currently, Synergeyes is the sole producer of hybrid lenses, offering products such as Duette, UltraHealth, and ClearKone, as listed in Table 4.5. These hybrid lenses offer multiple benefits, including enhanced comfort, improved visual clarity, and more stable vision. For further reading on hybrid lenses, refer to Sect. 3.2.2.

The contact lenses market offers a diverse selection of products from prominent manufacturers, with a particular focus on hydrogel and silicone hydrogel materials, designed with distinct features, especially for higher oxygen permeability (60–140 Dk). These companies produce various types of contact lenses, ranging conventional hydrogels and multiple generations of silicone hydrogel (SiHy) lenses, each with unique properties tailored to different needs. The key players in the contact lens market, including Alcon Vision Care, Bausch & Lomb, J&J Vision Care, Cooper Vision, Menicon, Contamac, Sauflon, and Paragon Vision Science, dominate the landscape. Tables 4.6 and 4.7 provide detailed comparisons of conventional hydrogel and multiple generations of silicone hydrogel contact lenses, respectively.

Oxygen permeability in a conventional hydrogel soft contact lens is directly related to its hydration degree [42]. When a soft contact lens is made from material with 100% water, its Dk value is similar to the of water, approximately 80×10^{-11} (cm^2/s) (mL O_2(mL $\times$ mmHg) [43]. This is due to the permeation of oxygen in the hydrogel contact lenses occurring through water. In contrast, silicone hydrogel contact lenses possess higher oxygen permeability due to the tris(trimethylsiloxysilyl)-γ-methacryloxy-propylsilane (TRIS) monomer units in their structure [44]. These lenses contain smaller pores in the silicone phase, which facilitate oxygen passage through the material. As demonstrated in Table 4.10, contact lens materials with higher oxygen permeability often exhibit lower EWC. For instance, Lotrafilcon A, used in Air Optix Night & Day lenses by Alcon Vision Care, has a high oxygen permeability (140 Dk) but lower EWC (24%). On the other hand, Galyfilcon A, used in Acuvue Advance lenses by J&J Vision Care, has lower oxygen permeability (60 Dk) but higher EWC (47%).

The comparison of various generations of hydrogel contact lens materials was summarized by Ishihara et al. [43], which highlights conventional hydrogel contact lenses, alongside the first, second, third, and fourth generations of silicon hydrogel contact lenses. The relationship of oxygen permeability and EWC of the lenses are summarized and illustrated in Fig. 4.4. Conventional hydrogel contact lenses typically show a wide range of EWC (approximately 34 –73%), but have low oxygen permeability (10–40 Dk). Meanwhile, the first and second generations of silicon hydrogel contact lenses demonstrate significant improvements in oxygen permeability, ranging from 60 to 140 Dk. Examples include Lotrafilcon B, Balafilcon A and Galyfilcon A from the first generation, and Senofilcon A, Comfilcon A and Asmoficon A from the second generation, all of which exhibit much higher oxygen permeability compared to conventional hydrogels. However, the lower EWC of the silicone hydrogel lenses observed in silicone hydrogels lenses is due to the hydrophobic nature of their components, such as silicone, TRIS and perfluoro moieties. Thus, to obtain a wettable surface, these lenses are often finished with surface treatments such as plasma oxidation for Balafilcon A and plasma polymerization of trimethylsilane, in the presence of oxygen (air) and methane for lotrafilcon A [45, 46]. When comparing the different generations of silicone hydrogel contact lenses, the third generation lenses show a notable improvement in EWC (38–48%) compared to the first and second generation (24–47%). This indicates that recent generations of silicone hydrogel contact lenses have enhanced EWC while retaining oxygen permeability comparable to earlier generations.

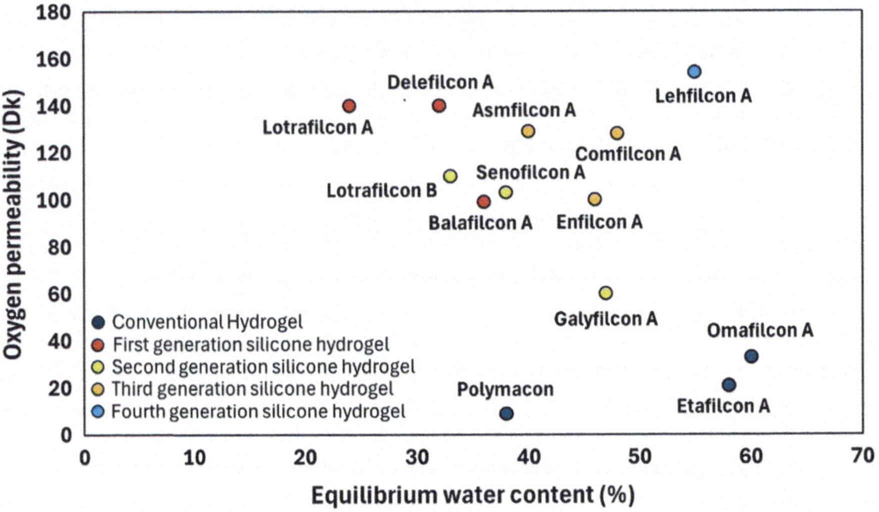

Fig. 4.4 Relationship between equilibrium water content and oxygen permeation of the commercially available contact lenses

4.4 Pioneering Research in Contact Lens Properties

Advancements in contact lens technology have significantly enhanced comfort, performance, and long-term wearability. Given that the ocular environment can directly impact the physical properties of contact lenses, this has prompted researchers to continue the explorations and strategies to optimize these interactions. This section covers recent research efforts aimed at advancing contact lens performance, which is organized into three main themes, as illustrated in Fig. 4.5. The first two themes present strategies to improve essential properties of contact lenses, focusing on **enhancing oxygen permeability and moisture retention** and **improving wettability for comfort and adaptability**. The third theme highlights **emerging innovations** in contact lenses for **drug delivery and smart lenses**, representing the next frontier in contact lens technology.

The ability of contact lenses to allow sufficient oxygen permeability and retain moisture is essential for wearer comfort and corneal health, making these properties a core focus in pioneering contact lens research. Figure 4.5 illustrates these advanced strategies as part of the broader goals of improving contact lens functionality for extended wear.

One promising strategy involves modifying silicone macromers to improve oxygen flow through the lens. By creating symmetric and asymmetric methacrylate-functionalized silicone macromers, Wuchte et al. [47] minimized phase separation between the hydrophilic and silicone components, thus enhancing lens stability. Maintaining a hydrophobic/hydrophilic ratio below 1.4 allowed the lenses to achieve a high oxygen permeability (Dk of 112) while retaining enough water content

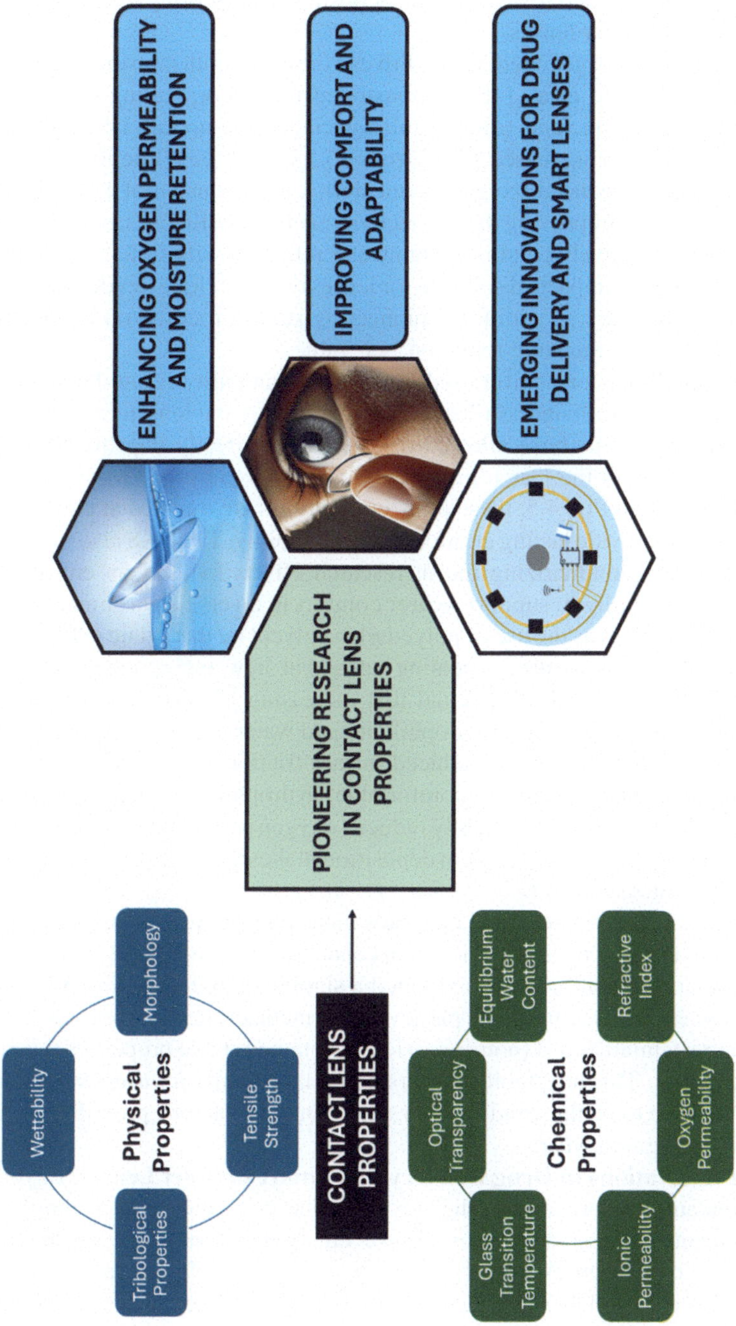

Fig. 4.5 Overview of pioneering advancements in contact lens technology which covers three themes in recent research: enhancing essential properties, improving comfort and adaptability, and emerging innovations

for comfortable wear. Importantly, the addition of fluorinated functionality further increased optical clarity and resistance to deposits, underscoring the strategy's potential for extended-wear lenses.

Another approach leverages chemical additives to enhance both moisture retention and oxygen transport. Liu et al. [48] demonstrated that incorporating carboxylic acid (acrylic acid) and surfactants (dioctyl sulfosuccinate sodium salt, DSS) into the lens material significantly increased EWC from 36.2 to 64.8%. While the addition of acrylic acid initially reduced oxygen permeability, the inclusion of DSS helped maintain oxygen flow by improving the connectivity within the silicone network. This balanced approach effectively combines high water retention with sufficient oxygen permeability, offering an impactful solution for lenses designed for extended comfort and health. For a summary of additional pioneering studies focused on enhancing oxygen permeability and moisture retention, see Table 4.8.

Enhancing wettability for comfort and long-term wearability is another critical factor in contact lens performance, influencing both comfort and long-term usability. Advances in wetting agents have allowed researchers to directly improve the lens surface's interaction with the tear film, a key factor for users prone to dryness or irritation. As depicted in Fig. 4.5, these pioneering efforts are directed at enhancing surface properties to provide lasting comfort and adaptability. Table 4.8 also includes studies in this theme, summarizing recent research efforts focused on enhancing wettability and adaptability to support wearer comfort in diverse environments.

One successful strategy is the use of polyethylene glycol methacrylate (PEGMA), traditionally employed as a surface coating agent but now incorporated into the lens material itself. Tran et al. [55] found that increasing PEGMA content in the lens significantly improved surface hydrophilicity and water retention, raising EWC from 31 to 55.4%. This enhancement reduced surface friction, indicated by a drop in the contact angle, resulting in a more comfortable, hydrophilic surface suitable for extended wear. Although PEGMA slightly reduced oxygen permeability, its contribution to lens flexibility and reduced protein deposition makes it a promising innovation for lenses that prioritize comfort.

For adaptive comfort, pH-sensitive materials have emerged as a novel approach, allowing lenses to respond to environmental pH changes. Ahn and Choi [56] demonstrated that etafilcon A (ionic) lenses exhibited a significant reduction in EWC and lens diameter when exposed to lower pH levels, as methacrylic acid becomes less ionized. In contrast, hilafilcon B (non-ionic) lenses retained stable properties regardless of pH variations. This adaptability offers potential benefits for users with fluctuating pH levels due to ocular conditions or environmental factors, providing more stable comfort throughout wear.

Emerging Innovations in Drug Delivery and Smart Contact Lenses. Beyond optimizing fundamental properties, the next frontier in contact lens technology involves therapeutic and smart lenses. These emerging innovations extend the functionality of contact lenses, transforming them from simple vision correction devices to multifunctional healthcare tools. Figure 4.5 outlines these pioneering advancements, highlighting the integration of drug delivery and health-monitoring capabilities in next-generation lenses.

Table 4.8 Summary of research studies in contact lens innovation

Authors	Manipulated strategy/Findings
Improving oxygen permeability and moisture retention	
Wu et al. [49]	• Developed silicone hydrogels using single/double methacrylate-terminated macromonomers • Longer polysiloxane chains increased oxygen permeability and EWC, while higher crosslinking reduced both
Sun et al. [50]	• Investigation on the effect of *N*-vinyl-pyrrolidone (NVP) and TRIS on the performance of conventional poly(methyl methacrylate) (PMMA) rigid contact lenses • Optimized methyl methacrylate (MMA)/NVP ratio (1:3) and TRIS content (60 wt.%) for PMMA lenses • Lenses achieved 95% light transmission, 27.8 Dk/t, and 18.7% EWC
Enhancing wettability for comfort and long-term	
Yamasaki et al. [51]	• Evaluated hyaluronic acid derivative (HAD) treatment for soft contact lenses (Etafilcon A, Senofilcon A, Comfilcon A, Lotrafilcon B) • HAD treatment improved wettability, decreasing contact angles (e.g., Etafilcon A: 86.0°, Senofilcon A: 77.5°)
Demian et al. [52]	• Developed super-hydrophilic coating for silicone hydrogels using tannic acid, dopamine, and chitosan derivative • Coated lenses achieved high hydrophilicity (5° contact angle) and lipid repellency without cytotoxicity
Emerging innovations in contact lenses	
Li et al. [53]	• Developed pHEMA hydrogels with cross-linked β-cyclodextrin- hyaluronan (β-CD-cHA) for tear protein resistance and sustained drug delivery • Hydrogels exhibited improved hydrophilicity (52° contact angle), oxygen permeability (36.4 Barrers), flexibility (1.8 MPa modulus), and 90% light transmittance, with effective conjunctivitis treatment in rabbits after 72 h
Li et al. [54]	• Investigated gold nanoparticles (GNPs) for enhanced bimatoprost drug uptake and controlled release • GNP-incorporated lenses showed sustained release up to 72 h, reduced burst release, and improved properties (67.6 Barrer Dk, 94.7% swelling, 98.6% transmittance)

This table compiles recent pioneering studies across three main themes: improving essential properties (oxygen permeability, moisture retention, and wettability) and exploring emerging applications in therapeutic and smart lenses

One notable advancement is in drug delivery lenses, which offer controlled and sustained medication release for ocular treatments. Maulvi et al. [57] incorporated Pluronic® F-68 to load hydrophobic drugs like gatifloxacin into contact lenses, achieving high optical transmittance and sustained release for up to 72 h. Further work by Wei et al. [58] introduced a circular-ring embedded silicone hydrogel with Eudragit® S100, which maintained high oxygen permeability and extended drug release for glaucoma treatment over 72 h. These lenses not only increase drug bioavailability but also minimize side effects by reducing the need for frequent eye drops.

Finally, smart contact lenses represent a pioneering direction in integrating real-time health monitoring into wearable devices. Keum et al. [59] developed a smart lens capable of monitoring glucose levels in tear fluid, a non-invasive alternative for diabetic patients. Similarly, Elsherif et al. [60] created a bifocal smart lens capable of detecting glucose concentration changes and aiding presbyopia management. These innovations provide non-invasive solutions for managing chronic health conditions, marking a significant step forward in contact lens technology. For an overview of additional studies in drug delivery and smart lens innovations, refer to Table 4.8.

Moreover, researchers are focusing on improving multifocal lenses to treat presbyopia. Kim et al. [61] has conducted a study on the efficacy of a new refractive multifocal contact lens for presbyopia. The multifocal lens used was OptaCon ZOOM (INTEROJO Co. Ltd. Pyeongtaek-si, Gyeonggi-do, Korea), with the primary endpoints were the corrected distance visual acuity (CDVA) and distance-corrected near visual acuity (DCNVA). From the results obtained from 22 patients with presbyopia, the DCNVA showed significant improvement after 1 week of wearing the contact lens (from 0.19 to 0.04 logMAR). Plus, OptaCon ZOOM received satisfaction scores of 96.2, 91.9, and 85.0 out of 100 from the patients. Therefore, this study suggested that the OptaCon ZOOM multifocal contact lens is effective for the correction of presbyopia.

In summary, the first section of this chapter elaborated on the international standard ISO 18369, which outlines the criteria and properties of contact lenses, covering rigid corneal, scleral, and soft contact lenses. It is divided into four parts, with ISO 18369-1 defining vocabulary terms and ISO 18369-2 specifying tolerance limits. Meanwhile, ISO 18369-3 describes the measuring methods for physical and optical properties, and ISO 18369-4 outlines the detailed measurement methods for physicochemical properties. Furthermore, the standard explains the naming of contact lens materials according to their properties, such as material composition and oxygen permeability. The next section elaborates on the essential characteristics of contact lenses, including lens metrology and their key properties. From the lens properties, the comparison among the commercially available contact lenses in the market is discussed, where the amount of silicone monomer units influences the oxygen permeability of the lenses. Moreover, pioneering research in contact lens properties emphasizes the advancements in therapeutic and smart contact lenses for health monitoring, where the affected properties were highlighted for further improvement.

References

1. The International Organization for Standardization: Ophthalmic optics—contact lenses—Part 3: Measurement methods (ISO 18369-3: 2017) (2017)
2. The International Organization for Standardization: Ophthalmic optics—contact lenses—Part 4: Physicochemical properties of contact lens materials (ISO 18369-4:2017) (2017)
3. Drzezo: Rigid contact lenses: basics. https://entokey.com/rigid-contact-lenses-basics/

4. Efron, N., Maldonado-Codina, C.: Development of contact lenses from a biomaterial point of view–materials, manufacture, and clinical application. In: Comprehensive Biomaterials, pp. 517–541. Elsevier BV (2011). https://doi.org/10.1016/B978-0-08-055294-1.00270-1
5. Lira, M., Santos, L., Azeredo, J., Yebra-Pimentel, E., Real Oliveira, M.E.C.D.: The effect of lens wear on refractive index of conventional hydrogel and silicone-hydrogel contact lenses: a comparative study. Contact Lens Anterior Eye **31**, 89–94 (2008). https://doi.org/10.1016/j.clae.2007.09.001
6. Varikooty, J., Keir, N., Woods, C.A., Fonn, D.: Measurement of the refractive index of soft contact lenses during wear. Eye Contact Lens **36** (2010). https://doi.org/10.1097/ICL.0b013e3181c8135f
7. Malakooti, N., Alexander, C., Alvarez-Lorenzo, C.: Imprinted contact lenses for sustained release of Polymyxin B and related antimicrobial peptides. J. Pharm. Sci. **104**, 3386–3394 (2015). https://doi.org/10.1002/jps.24537
8. Young, M.D., Benjamin, W.J.: Calibrated oxygen permeability of 35 conventional hydrogel materials and correlation with water content. Eye Contact Lens **29**, 126–133 (2003). https://doi.org/10.1097/01.ICL.0000062463.64717.86
9. Bhamra, T.S., Tighe, B.J.: Mechanical properties of contact lenses: the contribution of measurement techniques and clinical feedback to 50 years of materials development. Contact Lens Anterior Eye. **40**, 70–81 (2017). https://doi.org/10.1016/j.clae.2016.11.005
10. Horst, C.R., Brodland, B., Jones, L.W., Brodland, G.W.: Measuring the modulus of silicone hydrogel contact lenses. Optom. Vis. Sci. **89** (2012). https://doi.org/10.1097/OPX.0b013e3182691454
11. Jones, L., Dumbleton, K.: Soft contact lens fitting. In: Contact Lenses, pp. 207–222. Elsevier (2019). https://doi.org/10.1016/B978-0-7020-7168-3.00010-6
12. Fuentes, R., Fernández, E., Pascual, I., García, C.: UV-visible transmittance of silicone-hydrogel contact lenses measured with a fiber optic spectrometer. In: Proceedings SPIE, p. 8785AZ (2013)
13. Compan, V., Garrido, J., Manzanares, J.A., Andres, J., Esteve, J.S., López, M.L.: True and apparent oxygen permeabilities of contact lenses. Optom. Vis. Sci. **69** (1992)
14. Tran, N.P.D., Yang, M.C., Tran-Nguyen, P.L.: Evaluation of silicone hydrogel contact lenses based on poly(dimethylsiloxane) dialkanol and hydrophilic polymers. Coll. Surf. B Biointerfaces. **206**, 111957 (2021). https://doi.org/10.1016/J.COLSURFB.2021.111957
15. Orsborn, G.N., Zantos, S.G.: Corneal desiccation staining with thin high water content contact lenses. https://journals.lww.com/claojournal/Abstract/1988/04000/Corneal_Desiccation_Staining_With_Thin_High_Water.6.aspx
16. Jones, L.: Modern contact lens materials: A clinical performance update. Contact Lens Spectr. **17**, 24–35 (2002)
17. Watts, A., Colby, K.: Contact lenses for keratoconus. In: Alió, J.L. (ed.) Keratoconus: Recent Advances in Diagnosis and Treatment, pp. 187–194. Springer International Publishing, Cham (2017). https://doi.org/10.1007/978-3-319-43881-8_16
18. Optician: Guide to hybrid contact lenses. https://www.opticianonline.net/cpd-archive/4659/ (2024)
19. SynergEyes: What is the difference between ClearKone and UltraHealth? https://www.synergeyes.co.uk/PROFESSIONAL/hybrid-lenses/other-lenses/faqs-other-lenses/what-is-the-difference-between-clearkone-and-ultrahealth
20. Bausch, Lomb: Professional fitting guide: ALDEN OPTICAL® Soft (hydrophilic) contact lens for daily wear. https://www.bauschsvp.com/siteassets/pdf/Alden-Optical-Soft-Contact-Lens-FG.pdf (2022)
21. Bausch, Lomb: Package insert/fitting guide: Softlens® 38. https://ecp.bauschcontactlenses.com/siteassets/pdf/soflens-38-package-insert.pdf/ (2021)
22. Bausch, Lomb: Package insert/fitting guide: PureVision® 2. https://pi.bausch.com/globalassets/pdf/PackageInserts/Vision-Care/lenses/PureVision-2-PIFG.pdf (2019)
23. Bausch, Lomb: Package insert/fitting guide: Soflens® Toric. https://pi.bausch.com/globalassets/pdf/PackageInserts/Vision-Care/lenses/SofLens-Toric-PIFG.pdf (2019)

24. Bausch, Lomb: Package insert/fitting guide: Soflens® Daily disposable. https://ecp.bauschcontactlenses.com/siteassets/pdf/soflens-daily-disposable-package-insert.pdf/ (2019)
25. Bausch, Lomb: Package insert/fitting guide: Biotrue® ONEday. https://www.bausch.com/contentassets/9630b79cbef1474490a249b85c04682e/biotrue-oneday-package-insert.pdf (2020)
26. Alcon: DAILIES® AquaComfort PLUS® contact lenses product information. https://www.myalcon.com/professional/contact-lenses/daily/dailies-aquacomfort-plus/
27. CooperVision: Proclear® 1 day (daily). https://coopervision.com.sg/practitioner/our-products/proclear-family/proclear-1-day-daily
28. CooperVision: Proclear® multifocal. https://coopervision.com/practitioner/our-products/proclear-family/proclear-multifocal
29. CooperVision: Biomedics Toric (ocufilcon D) package insert. https://coopervision.com/sites/coopervision.com/files/pi01093_rev_b_biomedics_toric.pdf (2020)
30. Dutta, D., Willcox, M.D.: A laboratory assessment of factors that affect bacterial adhesion to contact lenses (2013). https://doi.org/10.3390/biology2041268
31. Chaudhary, S., Ghimire, D., Basu, S., Agrawal, V., Jacobs, D.S., Shanbhag, S.S.: Contact lenses in dry eye disease and associated ocular surface disorders. Indian J. Ophthalmol. **71**, 1142–1153 (2023). https://doi.org/10.4103/ijo.Ijo_2778_22
32. González-Méijome, J.M., González-Pérez, J., Fernandes, P.R.B., Ferreira, D.P.L., Mollá, S., Compañ-Moreno, V.: Silicone hydrogel materials for contact lens applications (2014)
33. Kim, E., Ehrmann, K.: Refractive index of soft contact lens materials measured in packaging solution and standard phosphate buffered saline and the effect on back vertex power calculation. Contact Lens Anterior Eye. **43**, 123–129 (2020). https://doi.org/10.1016/j.clae.2019.12.005
34. Holden, B.A., Stephenson, A., Stretton, S., Sankaridurg, P.R., O'Hare, N., Jalbert, I., Sweeney, D.F.: Superior epithelial arcuate lesions with soft contact lens wear. Optom. Vis. Sci. **78** (2001)
35. Nicolson, P.C., Vogt, J.: Soft contact lens polymers: an evolution. Biomaterials **22**, 3273–3283 (2001). https://doi.org/10.1016/S0142-9612(01)00165-X
36. Tranoudis, I., Efron, N.: Tensile properties of soft contact lens materials. Contact Lens Anterior Eye. **27**, 177–191 (2004). https://doi.org/10.1016/J.CLAE.2004.08.002
37. Sterner, O., Aeschlimann, R., Zürcher, S., Scales, C., Riederer, D., Spencer, N.D., Tosatti, S.G.P.: Tribological classification of contact lenses: from coefficient of friction to sliding work. Tribol. Lett. **63**, 9 (2016). https://doi.org/10.1007/s11249-016-0696-5
38. Mann, A., Tighe, B.: Contact lens interactions with the tear film. Exp. Eye Res. **117**, 88–98 (2013). https://doi.org/10.1016/j.exer.2013.07.013
39. Muntz, A., Subbaraman, L.N., Sorbara, L., Jones, L.: Tear exchange and contact lenses: a review. J. Optom. **8**, 2–11 (2015). https://doi.org/10.1016/j.optom.2014.12.001
40. Shaw, A.J., Collins, M.J., Davis, B.A., Carney, L.G.: Eyelid pressure and contact with the ocular surface. Invest. Ophthalmol. Vis. Sci. **51**, 1911–1917 (2010). https://doi.org/10.1167/iovs.09-4090
41. Moreddu, R., Vigolo, D., Yetisen, A.K.: Contact lens technology: from fundamentals to applications. Adv. Healthc. Mater. **8**, 1900368 (2019). https://doi.org/10.1002/adhm.201900368
42. Efron, N., Morgan, P.B., Cameron, I.D., Brennan, N.A., Goodwin, M.: Oxygen permeability and water content of silicone hydrogel contact lens materials. Optom. Vis. Sci. **84** (2007). https://doi.org/10.1097/OPX.0b013e31804375ed
43. Ishihara, K., Shi, X., Fukazawa, K., Yamaoka, T., Yao, G., Wu, J.Y.: Biomimetic-engineered silicone hydrogel contact lens materials. ACS Appl. Bio Mater. **6**, 3600–3616 (2023). https://doi.org/10.1021/acsabm.3c00296
44. Thissen, H., Gengenbach, T., du Toit, R., Sweeney, D.F., Kingshott, P., Griesser, H.J., Meagher, L.: Clinical observations of biofouling on PEO coated silicone hydrogel contact lenses. Biomaterials **31**, 5510–5519 (2010). https://doi.org/10.1016/j.biomaterials.2010.03.040
45. Tighe, B.: Silicone hydrogel materials: how do they work? Silicone hydrogels rebirth Contin. wear contact lenses, 1–21 (2000)
46. Weikart, C.M., Matsuzawa, Y., Winterton, L., Yasuda, H.K.: Evaluation of plasma polymer-coated contact lenses by electrochemical impedance spectroscopy. J. Biomed. Mater. Res. An

Off. J. Soc. Biomater. Japanese Soc. Biomater. Aust. Soc. Biomater. Korean Soc. Biomater. **54**, 597–607 (2001). https://doi.org/10.1002/1097-4636(20010315)54:4<597::AID-JBM170>3.0.CO;2-S
47. Wuchte, L., DiPasquale, S., Masterson, A., Vance, A., Goff, J., Arkles, B., Sulaiman, S., Byrne, M.: Characterization and analysis of extended-wear silicone hydrogel contact lenses utilizing novel silicone macromers. J. Biomed. Mater. Res. Part A. **110**, 1512–1523 (2022). https://doi.org/10.1002/jbm.a.37389
48. Liu, B.-T., Pan, W.-C., Lu, Y.-T., Sun, K.-T.: Enhancing oxygen permeability and water content in silicone hydrogels through carboxylic acid and surfactant incorporation. J. Taiwan Inst. Chem. Eng. **165**, 105745 (2024). https://doi.org/10.1016/j.jtice.2024.105745
49. Wu, B., Wiseman, M.E., Seitz, M.E., Tomić, K., Heise, A., Brougham, D.F., Litvinov, V.M.: Impact of morphology on O_2 permeability in silicone hydrogel membranes: new insights into domain percolation from experiments and simulations. J. Memb. Sci. **621**, 118970 (2021). https://doi.org/10.1016/J.MEMSCI.2020.118970
50. Sun, Z., Zhang, R., Sun, Y., Wang, X., Duan, L., Gao, G.: Methyl methacrylate rigid contact lenses doped with N-Vinylpyrrolidone and 3-[Tris(trimethylsiloxy)silyl] propyl methacrylate for improved oxygen permeability. Polymer (Guildf). **309**, 127439 (2024). https://doi.org/10.1016/j.polymer.2024.127439
51. Yamasaki, K., Drolle, E., Nakagawa, H., Hisamura, R., Ngo, W., Jones, L.: Impact of a low molecular weight hyaluronic acid derivative on contact lens wettability. Contact Lens Anterior Eye. **44** (2021). https://doi.org/10.1016/j.clae.2020.05.003
52. Demian, P., Nagaya, D., Refaei, R., Iwai, K., Hasegawa, D., Baba, M., Messersmith, P.B., Lamrani, M.: Enhancing performance of silicone hydrogel contact lenses with hydrophilic polyphenolic coatings (2024). https://doi.org/10.3390/jfb15110321
53. Li, R., Guan, X., Lin, X., Guan, P., Zhang, X., Rao, Z., Du, L., Zhao, J., Rong, J., Zhao, J.: Poly(2-hydroxyethyl methacrylate)/β-cyclodextrin-hyaluronan contact lens with tear protein adsorption resistance and sustained drug delivery for ophthalmic diseases. Acta Biomater. **110**, 105–118 (2020). https://doi.org/10.1016/j.actbio.2020.04.002
54. Li, Q., Ma, C., Ma, Y., Ma, Y., Mao, Y., Meng, Z.: Sustained bimatoprost release using gold nanoparticles laden contact lenses. J. Biomater. Sci. Polym. Ed. **32**, 1618–1634 (2021). https://doi.org/10.1080/09205063.2021.1927656
55. Tran, N.-P.-D., Yang, M.-C., Hasanah, N., Tran-Nguyen, P.L.: Effect of poly(ethylene glycol) methacrylate on the ophthalmic properties of silicone hydrogel contact lenses. Coll. Surf. B Biointerfaces. **217**, 112713 (2022). https://doi.org/10.1016/j.colsurfb.2022.112713
56. Ahn, J., Choi, M.: The pH-induced physical properties of ionic contact lens material. Heliyon. **9** (2023). https://doi.org/10.1016/j.heliyon.2023.e12996
57. Maulvi, F.A., Parmar, R.J., Desai, A.R., Desai, D.M., Shukla, M.R., Ranch, K.M., Shah, S.A., Shah, D.O.: Tailored gatifloxacin Pluronic F-68-loaded contact lens: Addressing the issue of transmittance and swelling. Int. J. Pharm. **581**, 119279 (2020). https://doi.org/10.1016/j.ijpharm.2020.119279
58. Wei, Y., Hu, Y., Shen, X., Zhang, X., Guan, J., Mao, S.: Design of circular-ring film embedded contact lens for improved compatibility and sustained ocular drug delivery. Eur. J. Pharm. Biopharm. **157**, 28–37 (2020). https://doi.org/10.1016/j.ejpb.2020.09.010
59. Keum, D.H., Kim, S.-K., Koo, J., Lee, G.-H., Jeon, C., Mok, J.W., Mun, B.H., Lee, K.J., Kamrani, E., Joo, C.-K., Shin, S., Sim, J.-Y., Myung, D., Yun, S.H., Bao, Z., Hahn, S.K.: Wireless smart contact lens for diabetic diagnosis and therapy. Sci. Adv. **6**, eaba3252 (2020). https://doi.org/10.1126/sciadv.aba3252
60. Elsherif, M., Alam, F., Salih, A.E., AlQattan, B., Yetisen, A.K., Butt, H.: Wearable bifocal contact lens for continual glucose monitoring integrated with smartphone readers. Small. **17**, 2102876 (2021). https://doi.org/10.1002/smll.202102876
61. Kim, D.Y., Ahn, H., Lee, S., Jun, I., Seo, K.Y., Kim, S.Y., Kim, T.-I.: Efficacy of a new refractive multifocal contact lens for presbyopia. Sci. Rep. **14**, 8878 (2024). https://doi.org/10.1038/s41598-024-55918-5

Chapter 5
The Industrial Landscape of Contact Lens Manufacturing

Abstract This chapter provides an in-depth exploration of the industrial processes and innovations that underpin the manufacturing of contact lenses. It begins with an overview of the journey from conceptual design to precision-engineered lenses, highlighting the critical factors that shape lens diameter, base curve, and optical zone for optimal performance. Key manufacturing techniques, including lathe-cutting, cast molding, and spin casting, are discussed with a focus on automation, precision, and scalability. The chapter also examines the challenges of quality control and defect prevention, emphasizing the importance of stringent standards and regulatory compliance to ensure product reliability. Advances in packaging and sterilization methods, such as aseptic filling and gamma radiation, are detailed as essential steps for maintaining lens sterility and user safety. A significant portion of the chapter is dedicated to recent technological innovations, including the integration of UV LED curing and advanced automation in lens production. These breakthroughs have revolutionized manufacturing efficiency and expanded the capabilities of contact lens design. By presenting a comprehensive view of the industrial landscape, this chapter underscores the intricate processes that drive the creation of high-quality contact lenses and highlights the continuous efforts to innovate and improve production methods.

Keywords Contact lens manufacturing · Cast molding · Spin casting · UV LED curing · Quality control · Lens defects · Automation · Lens packaging · Sterilization · Industrial innovation

5.1 From Concept to Precision Lens Creation

The creation of contact lenses has undergone a remarkable evolution, transitioning from handcrafted glass lenses to the high-precision, scalable manufacturing techniques of today. This transformation has been driven by advancements in lens geometry, materials, and production methods, which together ensure that lenses meet stringent optical, comfort, and safety standards. Understanding the interplay between

© The Author(s), under exclusive license to Springer Nature Singapore Pte Ltd. 2025

N. Adrus et al., *Contact Lenses*, Series in BioEngineering,
https://doi.org/10.1007/978-981-96-3636-5_5

key geometric parameters—lens diameter, base curve, and optical zone—and their impact on manufacturing processes is crucial for designing lenses that align with both user needs and production efficiency. These foundational design elements are discussed in Sect. 5.1.1, highlighting how they influence material handling, mold precision, and quality control. Section 5.1.2 delves into the development and significance of Wichterle's work, exploring how his rudimentary "lens-machine" laid the groundwork for scalable, automated production methods that continue to define the industry.

5.1.1 The Contact Lens Geometry

The essential design parameters of contact lenses—**lens diameter**, **base curve**, and **optical zone**—play a pivotal role in shaping manufacturing practices. These parameters directly influence tooling precision, material handling, and quality assurance measures, requiring manufacturers to strike a careful balance between functionality, user comfort, and production efficiency. This section explores how these interconnected design elements influence manufacturing adjustments needed to produce high-quality, defect-free lenses. A summary of these parameters and their technical design and manufacturing impacts is provided in Table 5.1.

Lens Diameter. The lens diameter is established through precise mold designs and adjustments to the lathe based on the size of the lens. For soft lenses, typical

Table 5.1 Summary of contact lens parameters towards the technical design and manufacturing impact

Parameter	Technical design impact	Manufacturing impact
Lens diameter	• Controls fit and centration; typical range for soft lenses is 13.8–14.5 mm • Larger diameters improve coverage but may increase thickness, affecting comfort	• Requires precision molds and tooling. Tolerance is ±0.1 mm for soft lenses and ±0.05 mm for RGP • Real-time laser systems verify diameter stability during processes like spin-casting and lathe-cutting
Base curve	• Determines fit and tear exchange; typical range is 8.0–9.0 mm for soft lenses, 7.0–8.5 mm for RGP lenses • Improper base curve affects comfort and oxygen permeability	• Requires precise lathe cutting or molding with tolerances of ±0.02 mm for RGP • Flexible materials like silicone hydrogel may need adjustments to retain the intended curve. Automated interferometry systems ensure curvature accuracy
Optical Zone	• Central vision zone: typically 6–8 mm in diameter to match pupil size and minimize aberrations • Multifocal lenses may have varying zone sizes	• Needs exact centering within ±0.05 mm. Advanced mold shaping and laser alignment are used, especially for multifocal and toric lenses • Coatings or finishes may be applied to enhance optical clarity

diameters range from 13.8 to 14.5 mm. This range ensures adequate overlap with the sclera, which provides stability and minimizes lens movement [1]. To maintain a consistent fit across different manufacturing batches, the diameter must be controlled within specific tolerances: ±0.1 mm for soft lenses and ±0.05 mm for rigid gas permeable (RGP) lenses (Table 5.1). RGP lenses require tighter tolerances due to their rigid structure. Larger scleral lenses, often exceeding 15 mm, are designed to vault over the cornea and rest on the sclera, providing enhanced stability and comfort, particularly for conditions such as keratoconus and severe dry eye. Proper control of these tolerances reduces the risk of lenses being too loose, which can cause excessive movement and discomfort, or too tight, which may restrict tear exchange and potentially harm corneal health [2].

Lens diameter also affects the optical zone, particularly in advanced designs like multifocal and toric lenses. Larger diameters can accommodate wider optical zones, improving vision correction for wearers with large pupils or multifocal needs. However, larger optical zones often require thicker lenses, which may reduce oxygen permeability and compromise comfort. Additionally, the peripheral edge profile, tied closely to diameter, requires thinning or beveling to minimize irritation and ensure smooth eyelid interaction during blinking. For specialized applications, such as scleral lenses, maintaining precise control over diameter is even more critical. These large-diameter lenses must vault over the cornea while resting comfortably on the sclera, ensuring proper tear flow and avoiding mechanical stress. To maintain dimensional accuracy, manufacturers use processes such as spin casting, cast molding, or lathe cutting (details about this process—see Sect. 5.1.2).

Figure 5.1 visually represents the differences in diameter among scleral lenses, soft lenses, and RGP lenses. Scleral lenses, with the largest diameter, are shown alongside soft lenses, which fall within the mid-range, and RGP lenses, which have the smallest diameter. This size progression reflects their respective design objectives: scleral lenses for stability and vaulting, soft lenses for balance between comfort and flexibility, and RGP lenses for precise vision correction with minimal corneal coverage.

Manufacturing lenses with precise diameters presents significant challenges. During polymerization or hydration, materials like silicone hydrogels undergo shrinkage, altering the final dimensions. To address this, manufacturers pre-adjust

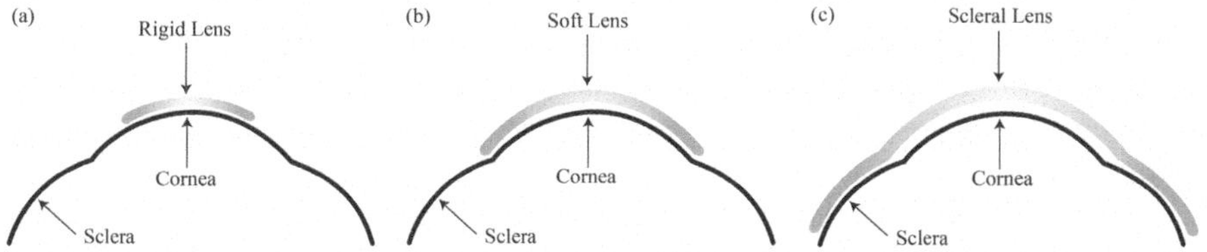

Fig. 5.1 Schematic illustration showing the difference in diameter among RGP, soft and scleral lenses. Scleral lenses have the largest diameter, vaulting over the cornea and resting on the sclera, while soft lenses and RGP lenses have progressively smaller diameters to meet different functional requirements

mold dimensions and **use laser-guided measurement** systems **and high-speed cameras** for real-time monitoring of lens geometry. These technologies detect deviations instantly, enabling immediate corrections and ensuring consistent quality. Larger diameters consume more raw materials, increasing production costs and waste. Optimized mold designs and precise dispensing systems help manufacturers mitigate these challenges while maintaining throughput.

Base curve. The base curve of the lens dictates the curvature settings in the molds or the paths of the lathe cutting process, which required a high degree of precision to achieve the desired sagittal depth [3]. This parameter is critical for ensuring comfort and effective vision correction. Typically, the range of base curves for soft contact lenses are from 8.3 to 9.0 mm, with an average of 8.6 mm. Steeper curves are ideal for individuals with rounded corneas, while flatter curves suit less pronounced shapes. Precise control of the base curve is essential to achieve the desired sagittal depth, as deviations can lead to poor lens stability or inadequate tear exchange.

For RGP lenses, achieving allowable tolerance requires high-precision molds or lathe paths to create curvature with tolerances as fine as ± 0.02 mm. To mitigate these issues, manufacturers utilize advanced automated systems equipped with real-time laser measurements to ensure that the optical zone is precisely centered, thereby reducing waste and maintaining high performance. Temperature control during polymerization is also critical for maintaining hydration levels and consistent curvature. By integrating advanced tooling and real-time monitoring systems, manufacturers can produce lenses with precise base curves that align with user needs and regulatory standards.

Optical Zone. The optical zone is the central area of the lens responsible for vision correction and typically ranges from 6 to 8 mm in diameter. The optical zone must be carefully calibrated to prevent visual distortions, particularly in low-light conditions when the pupil dilates. Misalignment of the optical zone can lead to spherical or peripheral aberrations, adversely affecting visual clarity [4]. To ensure precision, manufacturers employ real-time laser alignment systems and automated inspection technologies. These systems verify the centering and size of the optical zone, allowing immediate corrections during production to reduce material waste and maintain high performance. The optical zone must also be carefully integrated with other lens parameters, such as diameter and base curve, to create a seamless transition between functional and peripheral zones. For advanced lenses, such as multifocal and toric designs, the optical zone often requires additional finishing steps. Materials with unique refractive indexes, like silicone hydrogels, may need coatings to prevent glare and cloudiness, enhancing optical stability. These coatings are applied during post-production to maintain the lens's performance under varying lighting conditions.

Achieving Consistency Across Parameters. Consistency in these parameters across large production volumes is achieved through automated feedback systems and stringent quality control processes [5]. Laser-guided monitoring and automated inspection stations continuously verify essential parameters, triggering real-time adjustments whenever deviations are detected. Computer-controlled lathes and precisely calibrated molds ensure that lenses meet exact design tolerances, while real-time measurements allow for immediate corrections of any variations in diameter or base curve. **Advanced imaging technologies**, including **3D scanning of the eye**, allow manufacturers to produce scleral lenses tailored to individual corneal topographies [6]. This integration of advanced automation enables manufacturers to balance the need for large-scale production with customization, ensuring that lenses meet specific individual requirements without sacrificing production speed or quality. Additionally, precise machining processes are employed to achieve the necessary vault and edge profiles, ensuring a secure and comfortable fit.

In addition, specific manufacturing considerations are required for contact lenses with different properties such as flexibility and oxygen permeability. For example, higher flexibility of silicone hydrogels requires specialized mold release techniques to ensure dimensional stability and temperature control to maintain consistent hydration levels [7]. Moreover, the hydrophobic nature of silicone necessitates the use of specialized non-stick surfaces or coatings during molding to prevent adherence, to ensure the lens maintains the required diameter and base curve without deformation. Together, these material-specific adaptations allow manufacturers to produce comfortable lenses that meet oxygen transmission standards without sacrificing durability. Additionally, materials with unique refractive indexes, such as silicone hydrogel, may require specific finishing steps to maintain optical stability [8]. This might include applying coatings that prevent cloudiness or glare in lenses for multifocal and toric designs.

By integrating design parameters such as diameter, base curve, and optical zone with precision engineering, manufacturers uphold stringent quality standards and align each product with regulatory requirements and user expectations. Advanced technologies such as real-time laser-guided feedback systems, automated inspection stations, and tools like finite element analysis (FEA) enable manufacturers to maintain tight tolerances and reduce defects, even in high-volume production. These advancements not only enhance the scalability of lens production but also allow for the development of complex, personalized designs, meeting the growing demand for diverse and high-performance contact lenses.

5.1.2 Wichterle's Spin-Casting Invention: Setting the Stage for the Lens Manufacturing

The manufacturing process of contact lenses has evolved significantly from its early days of handcrafted glass lenses to the sophisticated techniques used today, evolving from handcrafted glass lenses to highly automated, precision-engineered systems. A key milestone in this evolution was Otto Wichterle's pioneering of the spin-casting method in 1961, which became the foundation for today's commercial-scale production of soft contact lenses.

Spin-Casting Method. Wichterle's work began with challenges in lens fabrication, where initial attempts to use polystyrene molds resulted in lenses with irregular edges. Drawing inspiration from everyday phenomena, Wichterle noticed how mixing sugar into coffee formed a smooth parabolic surface. This observation led him to hypothesize that a similar principle could be applied to lens shaping by spinning a liquid polymer. This idea became the foundation for the spin-casting process. In December 1961, using a children's Merkur construction kit, a bicycle dynamo, and a bell transformer, Wichterle built a rudimentary apparatus in his home (Fig. 5.2) [9]. Demonstrating extraordinary creativity, he also handcrafted molds and glass tubing to dose the molds with monomer. With this setup, humorously called the "lens-machine," Wichterle achieved a major breakthrough. On Christmas afternoon, assisted by his wife Linda, he successfully produced the first four hydrogel lenses. These lenses had smooth edges, a significant improvement over earlier attempts, and Wichterle tested them on his own eyes. Despite being the wrong optical power, the lenses were comfortable, proving the feasibility of his method.

The spin-casting process in principle involves depositing a liquid polymerization mixture into a mold to achieve the desired lens shape. As the mold rotated at a carefully calibrated number of revolutions per minute (rpm), the interplay of centrifugal force, surface tension, and the mold's inner contours shaped the liquid polymer into a lens with smooth, uniform edges. The centrifugal force pushed the polymer outward, creating the base curve, while the surface tension of the liquid ensured that the lens surface remained smooth and free from imperfections. The mold's inner geometry

Fig. 5.2 The rudimentary "lens-machine" constructed in 1961 using the Merkur kit, a bicycle dynamo, and a bell transformer to spin-cast hydrogel contact lenses (reused from Duskova-Smrckova et al. [9], license under Creative Commons Attribution 4.0)

further refined the curvature and thickness of the lens, ensuring optical precision. Once the polymerization process was completed, the hardened lenses were hydrated to transform them into soft hydrogel lenses with the flexibility and comfort required for wear. This hydration step was critical for imparting the lenses with their unique water-retentive properties, a hallmark of modern soft contact lenses.

A few days later, Wichterle submitted his patent application for the spin-casting process and went on to produce over 100 lenses with his apparatus. Determined to scale the process, he designed new prototypes, incorporating multiple spindles to increase efficiency. These improvements required more power, leading him to replace the dynamo with a stronger motor sourced from his gramophone. By the first four months of 1962, Wichterle and his team had produced an impressive 5500 lenses, marking the beginning of scalable soft contact lens production.

What made this process particularly innovative was its scalability and reproducibility. Unlike earlier manual methods, spin-casting allowed multiple lenses to be produced simultaneously with consistent quality. Wichterle's handcrafted molds and glass tubing enabled precise dosing of the liquid polymerization mixture, minimizing material waste and ensuring uniformity across lenses. This approach laid the groundwork for modern manufacturing techniques, proving that high-quality lenses could be produced efficiently on a larger scale.

5.2 Core Manufacturing Process

In this section, the primary manufacturing processes for contact lenses are divided into three main methods: **spin casting, lathe cutting, and cast molding**. Each method is meticulously designed to produce lenses with exact shapes, sizes, and optical properties tailored to the needs of users. Following the fabrication of contact lenses, Sect. 5.2.2 delves into the inspection process to eliminate defects and ensure the quality of the lenses, as well as the packaging and sterilization processes for hygiene and safety standards. These steps are essential for providing reliable and comfortable products to consumers. Recent innovations have further enhanced lens production by incorporating advanced technologies, such as automation and AI-driven quality control, thereby improving efficiency and precision in the manufacturing process, which are covered in Sect. 5.3.

5.2.1 Contact Lens Fabrication Methods

The manufacturing process of contact lenses has evolved significantly from its early days of handcrafted glass lenses to the sophisticated techniques used today, transitioning from rudimentary setups to highly automated, precision-engineered systems. Initially, lenses were made using spin-casting, pioneered by Otto Wichterle in 1961, which laid the foundation for today's commercial-scale manufacturing processes.

Modern production methods include **spin-casting, lathe-cutting,** and **cast molding**, each offering distinct advantages and being used for different types of contact lenses, such as soft hydrogel lenses and RGP lenses.

Spin-Casting Method. This method involves pouring the contact lens formulation into a mold, followed by spinning it at a high speed. The resulting centrifugal force distributes the contact lens formulation evenly across the surface of the mold, creating the desired lens shape [10]. Figure 5.3 shows the illustration of the spin casting process, from the creation of the mold to the packaging step. Initially, the female mold made of metal or glass is prepared, cleaned, and coated with a release agent using the male tool to form the desired contact lens shape [5]. Then, the contact lens formulation is dispensed into the mold and spun to shape the lens. The contact lens formulation is solidified into a solid lens structure through a curing process, such as exposure to UV light or thermal treatment. After the completion of the polymerization process, the lens is removed and further undergoes additional processes such as rinsing surface treatments, inspection and packaging. Interestingly, this technique offers advantages, which allowed for the mass production capabilities and consistency in lens manufacturing [11]. Furthermore, the mold used in spin casting method allows for the lens production with customized designs and complex geometries, and particularly advantageous for producing soft contact lenses.

Lathe-Cutting Method. The lathe-cutting method is a traditional, yet highly precise technique used in the manufacturing of contact lenses. This process begins with the selection of appropriate lens materials, typically supplied in the form of rods or buttons made from polymethyl methacrylate (PMMA) for rigid lenses or various hydrogels for soft lenses. Figure 5.4 shows an illustration summary of the lathe-cutting process. Firstly, these materials are mounted on a spinning shaft of a lathe, where they are shaped and polished into the desired lens curvature [12]. There are two cutting stages involved in the lathe cutting method, which are rough cutting and fine cutting. The rough cutting stage removed the excess material to achieve the basic shape of the lens. Then, fine cutting refines the lens curvature and surface quality using finer tools and adjusts the prescription and design specifications. Subsequently, the lens is further continued for polishing to enhance the surface smoothness and optical clarity, as well as eliminating imperfections and ensuring comfort. Plus, edge finishing is carried out to ensure the lens edges are smooth for a comfortable fit by employing specialized tools to create rounded or bevelled edges. Lastly, the lenses are inspected to confirm if each lens meets the required specifications.

The lathe-cutting method offers several advantages, including high precision and the ability to produce custom lenses tailored to individual patient needs. This method is particularly suitable for creating lenses with complex designs, such as toric lenses for astigmatism or multifocal lenses for presbyopia. However, the process is more time-consuming compared to other methods like injection molding, as each lens must be individually shaped and polished. Furthermore, the process was entirely manual, requiring skilled technicians to shape and polish each lens by hand [5]. Despite these challenges, the precision, customization capabilities, and material versatility of lathe-cutting ensure it remains a vital technique in the contact lens industry. Advances in technology, such as computer-controlled lathes and automated inspection systems,

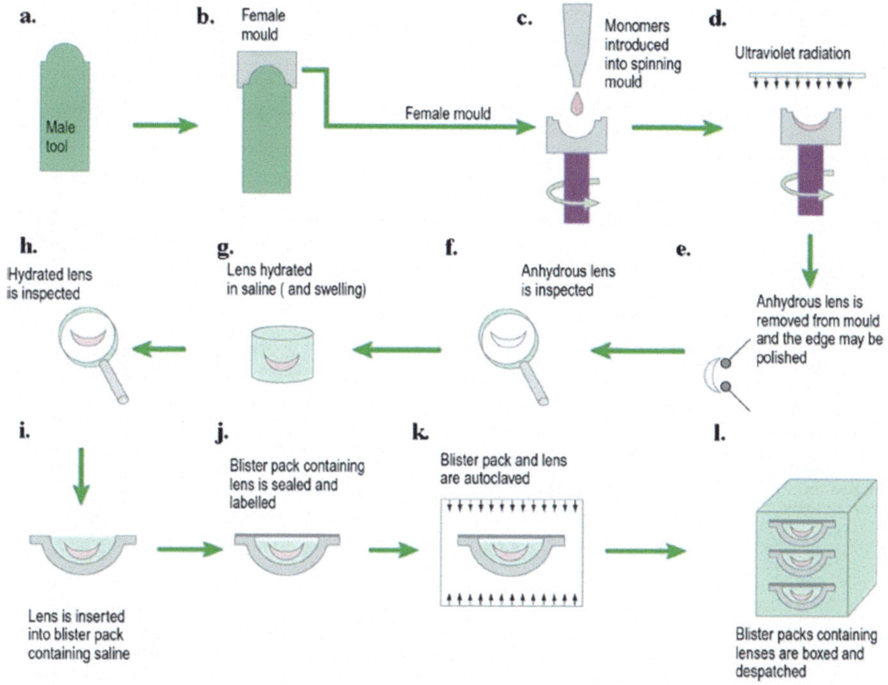

Fig. 5.3 Illustration of spin-casting method for contact lens manufacturing, from the mold preparation to the lens packaging (reused with permission from Shaker et al. [5], license under CC BY)

have further enhanced the efficiency and quality of lathe-cut lenses, making them a reliable choice for high-quality vision correction [13].

Cast Molding. The most popular method to produce contact lenses on a mass scale is due to multiple lenses can be produced in one process. This method consists of male and female mold sections that are usually made of the copolymer of propylene or ethylene [14]. Figure 5.5 illustrates the process of contact lens manufacturing via cast molding. Initially, the molds are produced via injection molding, and the male and female sections are assembled to form a cavity in the contact lens shape when combined. In the cast molding process, the contact lens formulation is dispensed on the female section (front curve mold) and covered by the male section (back curve mold). The excess contact lens formulation is squeezed out after the alignment of the male and female molds. Then, the lens formulation is cured until it solidifies to form a contact lens. The final step of the process is similar for inspection and packaging.

Remarkably, the cast molding method offers several significant advantages that make it the preferred technique for manufacturing contact lenses on a mass scale. It is highly efficient and scalable, enabling the production of large volumes of lenses at reduced per-unit costs, making it particularly suitable for disposable and daily wear lenses. The method ensures consistency in quality and precision, producing

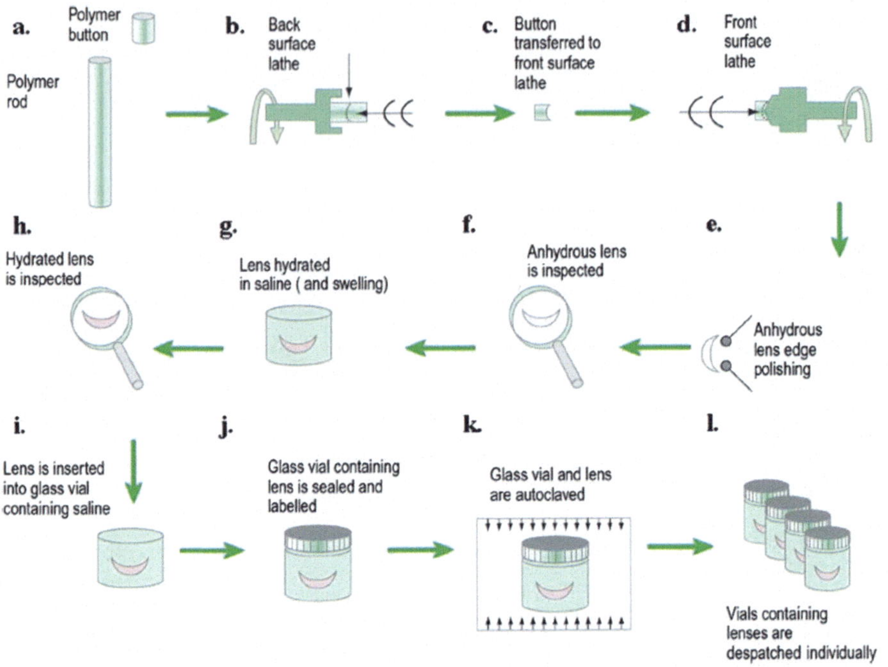

Fig. 5.4 Illustration of lathe-cutting method for contact lens manufacturing, from lathe spinning to the packaging step (reused with permission from Shaker et al. [5], license under CC BY)

lenses with uniform dimensions and optical properties. Additionally, cast molding is cost-effective due to its streamlined process and adaptability to various lens designs and materials. Its customization potential allows for the creation of lenses tailored to specific user requirements, including those made from advanced materials like silicone hydrogels for improved comfort and oxygen permeability. These advantages make cast molding the dominant choice for manufacturing soft contact lenses, especially disposable types, which are now the most widely used lenses globally. Table 5.2 summarizes the benefits and drawbacks for each process of contact lens fabrication.

5.2.2 Inspection, Packaging and Sterilization for Contact Lenses

The final stages of contact lens manufacturing encompass **inspection, packaging, and sterilization**, each playing a pivotal role in ensuring a safe, high-quality product reaches the consumer. During the inspection phase, lenses are subjected to rigorous

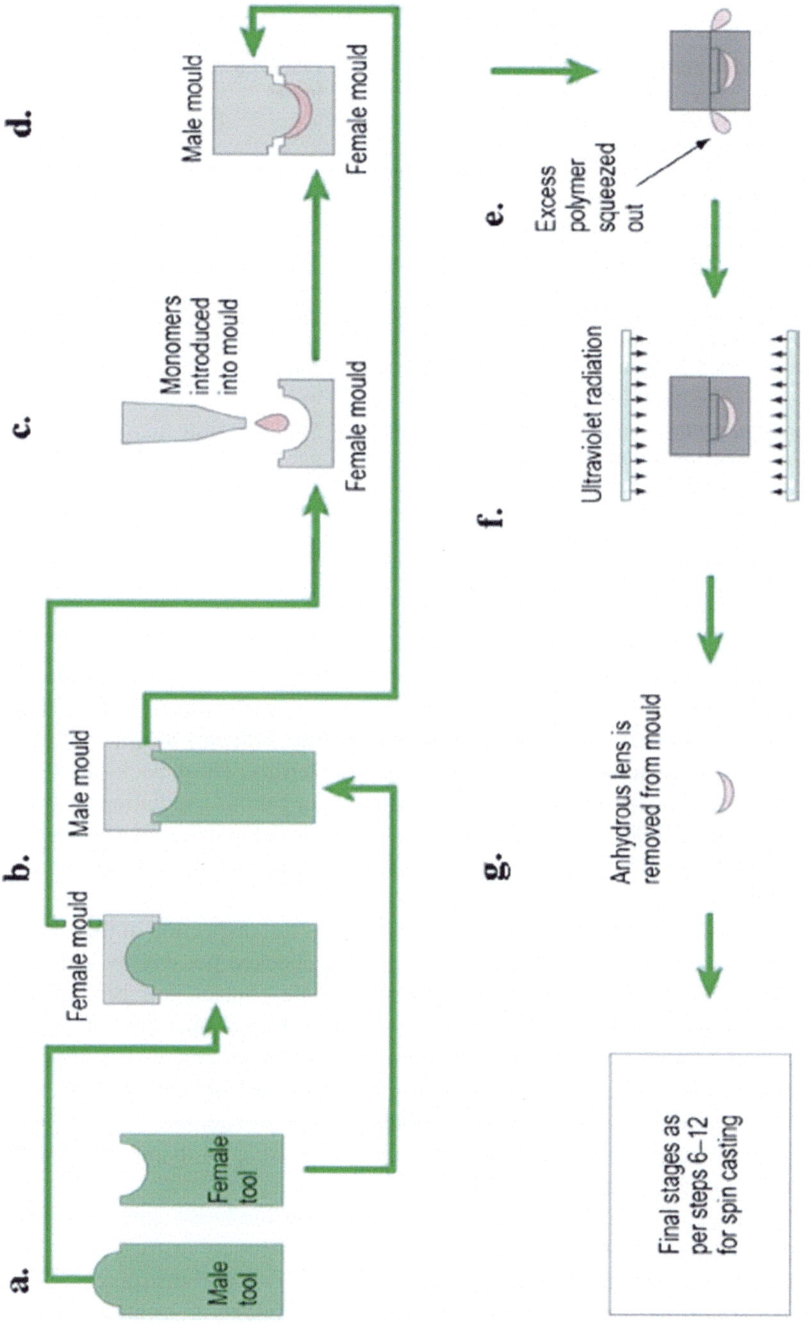

Fig. 5.5 Cast molding process in manufacturing contact lens, from mold preparation to the curing process of the contact lens (reused with permission from Shaker et al. [5], license under CC BY)

Table 5.2 Summary of different processes of contact lens fabrication with their benefits and drawbacks

Process	Benefits	Drawbacks
Spin casting	• Suitable for **mass production** with high efficiency • Ensures **consistency** in dimensions and optical properties • Allows for lenses with **customized designs** and **complex geometries** • Advantageous for **soft contact lenses**	• Limited flexibility for **rigid lenses**
Lathe cutting	• Offers **high precision**, ideal for custom and specialized lenses • Suitable for complex designs such as **toric** or **multifocal lenses** • Advances like **computer-controlled lathes** have improved efficiency and quality	• **Time-intensive** as lenses are individually shaped and polished • Labor-intensive • Higher production costs compared to mass-production techniques
Cast molding	• Highly **efficient and scalable**, ideal for **mass production** of soft lenses • Ensures **uniform quality** and optical precision in lenses • Adaptable to **various designs** and advanced materials like silicone hydrogels for added comfort	• **Material waste** from excess polymer during mold alignment • Unable to support gelation without the use of UV curing

quality control to verify they meet exacting standards for parameters such as diameter, base curve, and optical zone, as well as cosmetic specifications. Lenses that pass these stringent checks are **sealed in sterile saline solution as part of the primary packaging process**, ensuring hydration and sterility. These packs are then subjected to **sterilization processes** such as autoclaving or gamma irradiation to eliminate microorganisms. Once sterilized, the packs are enclosed in protective boxes for **secondary packaging**, which safeguards the lenses during transportation and storage while providing critical product information.

After the lenses have been through further processing, full inspection of the lenses is conducted, up to 4000 contact lenses were visually checked per day by a single inspector [15]. This 100% full inspection of large-scale contact lens production ensures that no lenses are skipped in the process. Well-trained operators are typically assigned to perform inspections, as they are skilled at accurately distinguishing between good and defective lenses. Some major defects found in contact lenses are summarized in Table 5.3, which include bubbles, scratches, roughness notch, and tear. Once the defective lenses are identified, they are set aside for disposal and will not be included in the packaging process.

Generally, inspection of the lenses is carried out via hardware approach. For instance, optical systems such as a common projected-based instrument (e.g., Optimec JCF) and an optical coherence tomography (OCT)-based instrument (e.g., Optimec is830) are the hardware approaches to detect defects in contact lenses. Among these two instruments, Coldrick et al. [18] have made a comparison, where

Table 5.3 Common types of defects found in contact lenses [16, 17]

Defects	Description of defects	References
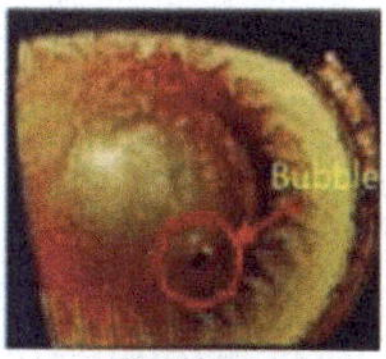 **Bubbles**	Bubbles are formed due to the air trapped under the contact lens	[16]
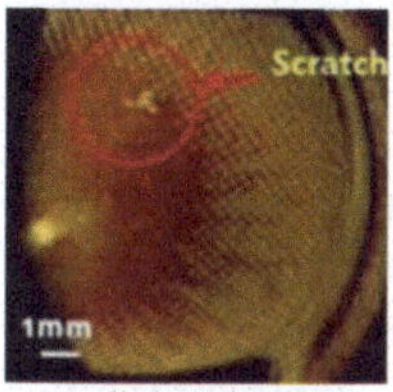 **Scratches**	A visible line or scratch on the lens surface	[16]
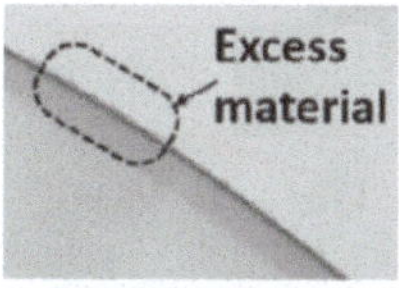 **Excess material**	Irregular or excess material on the edges or surfaces of the lens	[17]
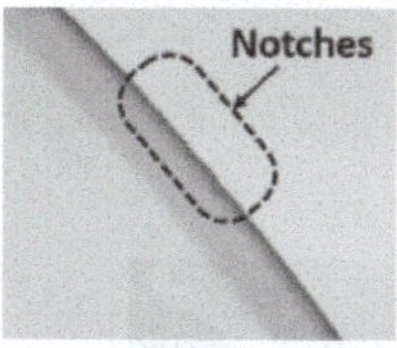 **Roughness notch**	A small indentation on the lens edge	[17]
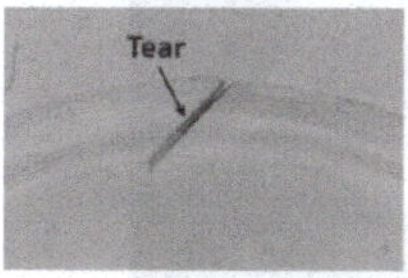 **Tear**	A split in the lens material	[17]

the OCT-based instrument is more reliable in measuring the geometric data at the center and edges of the lens. The difference between the images formed by an OCT-based instrument and a common projection-based instrument is illustrated in Fig. 5.6, where the edge defect can be clearly identified for OCT. Besides, using a 3D stereoscopic image rendering using OCT generates a tomographic image, which could not be seen with a magnifying projector instrument. As illustrated in Fig. 5.7, the scratch defect is clearly visualized, from the measured image in 3D. Hence, tomographic images produced through 3D stereoscopic rendering serve as an effective technique for detecting defects such as edge defects, bubbles, and scratches [16].

Inspection is an essential component in maintaining quality assurance within the contact lens manufacturing industry. Lenses that meet stringent inspection criteria are authorized for packaging and distribution to retailers. In this industry, it is imperative that all lenses undergo a 100% inspection process, as there is no allowance for defects. Consequently, conducting thorough inspections is a fundamental responsibility for key manufacturers prior to the sale of their products. Therefore, it is advisable for consumers to purchase only certified contact lenses to mitigate the risks associated with defective lenses, which may result in irritation and complications related to lens insertion and fitting.

Primary Packaging. Lenses that successfully pass quality inspection are then subjected to the packaging process, where primary packaging plays a crucial role in preserving sterility, hydration, and overall quality. To better understand the role of primary packaging in maintaining lens integrity, it is useful to compare the two

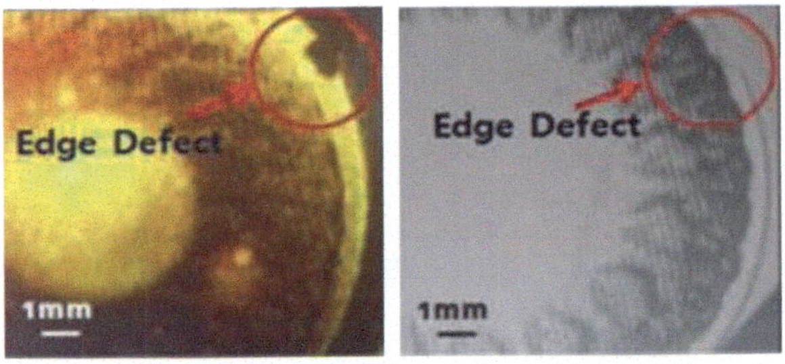

Fig. 5.6 The difference between the images formed for edge defect inspection by an OCT-based instrument (left) and a common projection-based instrument (right) [16]

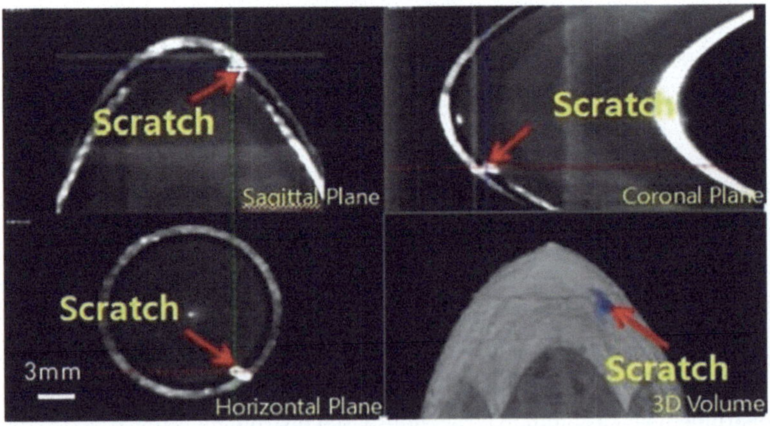

Fig. 5.7 Tomographic image formed for scratch defect via 3D stereoscopic rendering OCT [16]

primary packaging types—blister packs and glass vials—highlighting their unique features, advantages, and applications, which are outlined in Table 5.4.

As seen in Table 5.4, the most common primary packaging for disposable soft contact lenses is the blister pack, a lightweight plastic container filled with sterile saline and sealed with a vapor-barrier foil. Blister packs are cost-effective, easy to use, and protect lenses from damage, making them ideal for daily, weekly, or monthly disposables. In contrast, glass vials, sealed with a rubber bung and metal clip, are used for specialty lenses like scleral and RGP lenses. While more expensive, they provide superior sterility and durability, making them suitable for long-term storage and specialized applications. These packaging options cater to both disposable and specialty lens needs.

A contact lens blister package consists of **a base sheet** and **a sealing sheet** (Fig. 5.8). The base sheet, typically made from sturdy thermoplastic materials like polypropylene, is heated until they become pliable and then molded into the desired shape using a forming die [19]. The materials used for packaging must withstand sterilization processes, such as autoclaving, without degrading or compromising their barrier properties. For example, the plastic used in blister packs must resist warping or melting under high temperatures. Once the blister packs are formed, each cavity is filled with a sterile buffered saline solution, and the contact lenses are fully submerged within it to ensure hydration. This solution, with a neutral pH of 7.0–7.4, mimics

Table 5.4 Comparison of the two primary packaging types for contact lenses: blister packs and glass vials

Type of primary packaging	Features/Advantages	Applications
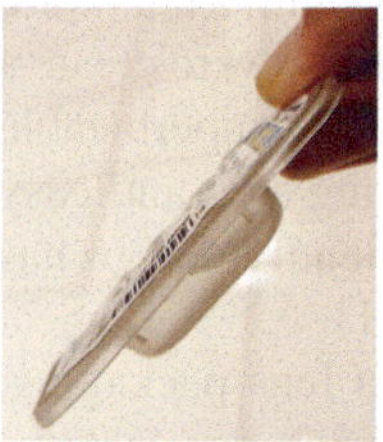 **Blister Pack**	Filled with saline solution, sealed with a foil lid • Lightweight • Cost-effective • Easy to open • Single-use design	Soft disposable contact lenses (daily, weekly, monthly)
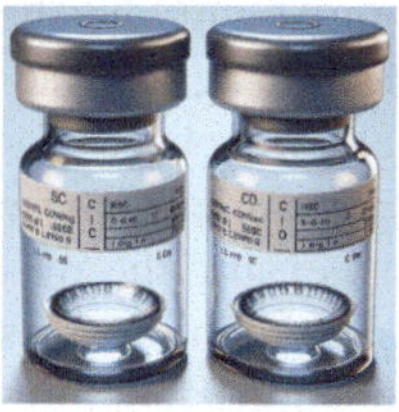 **Glass Vials**	Filled with saline solution, sealed with a rubber bung and metal clip • High sterility and durability • Suitable for long-term storage and specialized applications	Non-disposable soft contact lenses Specialty lenses (scleral, RGP)

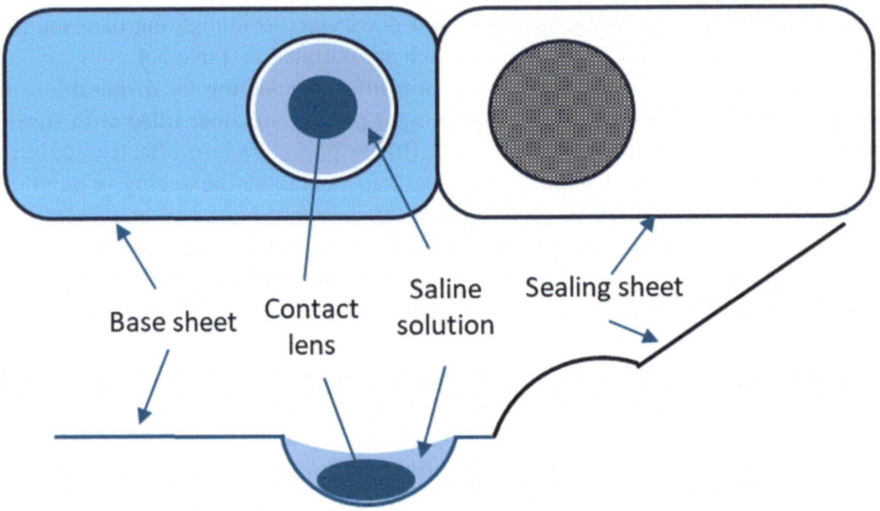

Fig. 5.8 Top and side views of a contact lens blister package, showing the base sheet with the lens cavity and the sealing sheet enclosing the lens and saline solution

human tears to prevent irritation and is buffered to maintain stability, preventing the pH from becoming acidic during storage [20].

The manufacturing process for **sealing blister pack** contact lenses is a meticulous and highly controlled procedure designed to ensure the sterility and safety of the lenses. The sealing sheet, usually made of aluminum foil, provides a vapor-barrier seal to ensure sterility and prevent contamination. It features a dome with a curvature radius of 13–19 mm and a height under 7 mm, which extends into the bowl to create a cavity that securely holds the lens. The heat-sealed packaging remains impermeable to bacteria and airtight, until the foil layer is peeled away, allowing safe and easy retrieval of the lens. This lightweight, cost-effective, and user-friendly design has made blister packs the standard for soft contact lens packaging [21].

Hamilton [22] outlines the critical features of effective contact lens packaging, emphasizing functionality and sustainability as summarized in Fig. 5.9. Key features include lens protection, optimized package size and weight, enhanced ease of lens removal, and structural integrity to ensure the lens remains fully submerged, aligned, and protected from inversion or misalignment. Additionally, the figure highlights the importance of reducing environmental impact through sustainable packaging practices. After sealing, blister packs undergo rigorous inspection, including checks for proper sealing, correct saline solution volume, and accurate lens positioning. These quality control processes, often performed by automated systems, ensure the reliability and consistency of the final packaged product [23]. The next step involves labelling the blister packs with essential information such as batch numbers, expiration dates, and product details. This is crucial for traceability and compliance with regulatory standards.

Fig. 5.9 Key features of contact lens packaging, focusing on lens protection, structural integrity, minimized packaging size and weight, environmental sustainability, ease of lens removal, and rigorous quality control

Sterilization Methods for Contact Lenses. Sterilization is a critical step in the manufacturing process of contact lenses. It ensures that any microorganisms present on the lenses or packaging are destroyed, making the lenses safe for use. Several sterilization methods are employed, each with distinct advantages and limitations (Fig. 5.10) [24]. These methods can be broadly categorized into **thermal**, **chemical**, and **radiation-based** approaches.

Thermal methods, such as autoclaving, are the most widely used for contact lenses, due to their effectiveness and reliability. Autoclaving involves exposing lenses and their packaging to high-pressure steam at 121–134 °C. This method effectively sterilizes both the lenses and their packaging in one process, as steam penetrates the packaging to eliminate microorganisms. The durability of materials like polypropylene, used in blister packs, ensures they withstand high temperatures without warping or degrading [25]. Precise pressure control is essential during autoclaving to prevent deformation or rupture of the blister packs, necessitating advanced autoclave systems that maintain consistent conditions throughout the process [26]. Additional thermal methods, such as dry heat and moist heat, are sometimes employed for specific lens types or materials.

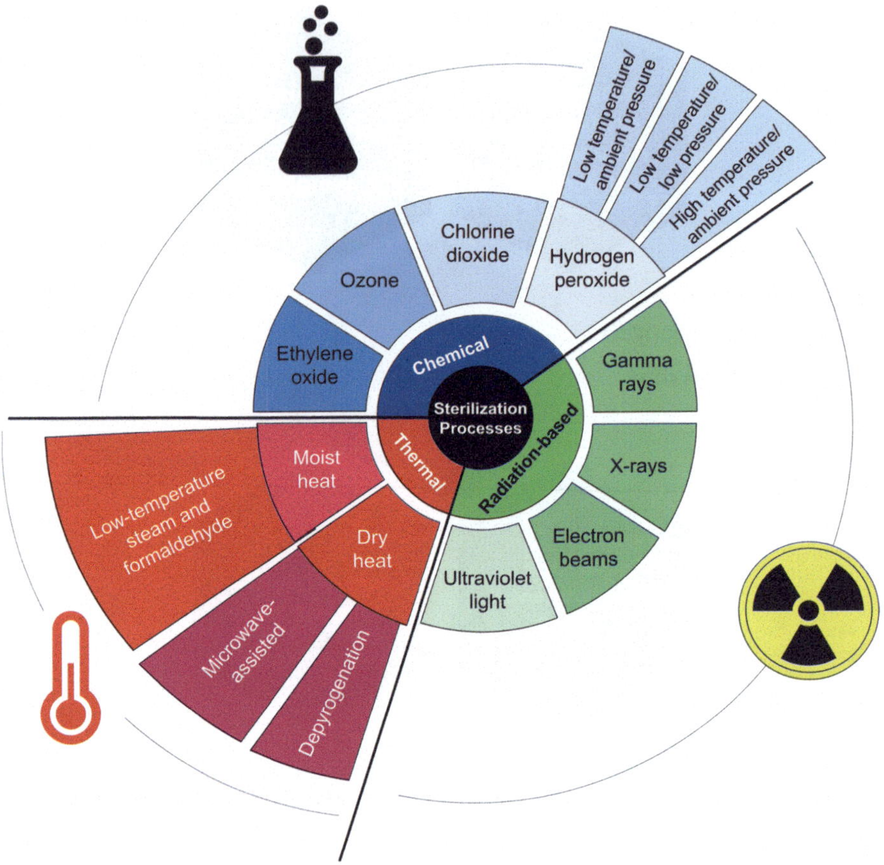

Fig. 5.10 Overview of sterilization processes for contact lenses, categorized into thermal, chemical, and radiation-based methods (reused with permission from Jildeh et al. [24])

Chemical sterilization methods are ideal for heat-sensitive lenses. Ethylene oxide (EtO) is the most commonly used chemical sterilant for such applications. EtO gas diffuses through packaging and surfaces, reacting with the nucleic acids and proteins of microorganisms to destroy them. This method is particularly suitable for sterilizing lenses made from heat-sensitive materials or housed in complex, porous packaging structures. However, due to its toxic and flammable nature, EtO requires specialized handling and extensive aeration to remove residual gas post-sterilization [27]. Other chemical agents, such as hydrogen peroxide, ozone, and chlorine dioxide, are also effective but may have limitations based on material compatibility.

Radiation-based methods, particularly **gamma radiation**, is another effective method for contact lenses that cannot tolerate high temperatures or chemicals. Gamma rays, typically emitted from a cobalt-60 source, penetrate deeply into the packaging, ensuring thorough sterilization without leaving chemical residues.

This method is particularly advantageous for maintaining the structural and functional integrity of hydrophilic lenses while ensuring sterility [28]. Radiation-based methods, including ultraviolet light and electron beams, are also used in specific contexts but may require specialized equipment and rigorous safety protocols.

Comparative Sterilization Insights and Importance. Autoclaving remains the preferred sterilization method for most contact lenses due to its efficiency and compatibility with commonly used materials like polypropylene. However, for lenses sensitive to heat or requiring specific packaging, EtO and gamma radiation offer viable alternatives. Gamma radiation's ability to sterilize at ambient temperatures makes it particularly effective for modern lens packaging, aligning with the FDA's premarket guidelines for daily-wear contact lenses. These methods ensure high sterility assurance levels without compromising the lenses' optical or physical properties [28].

Secondary Packaging and Final Stage of Manufacturing. Throughout the manufacturing process, maintaining a sterile environment is critical to ensure the safety and quality of contact lenses. Cleanroom conditions are maintained to prevent contamination, with all equipment and materials sterilized according to strict hygiene protocols. Automated systems play a vital role in minimizing human contact, thereby reducing contamination risks and enhancing process consistency. Packaging materials, such as high-density polyethylene (HDPE) and Tyvek®, are increasingly favored for their excellent microbial barrier properties, durability, and compatibility with various sterilization methods [27].

After sterilization, labeled blister packs are placed into **secondary packaging**, such as cartons, to provide additional protection during transportation and storage. This secondary packaging safeguards lenses against physical damage, temperature fluctuations, and environmental exposure, ensuring they remain intact throughout distribution. Secondary packaging also serves as a platform for labeling and branding, enabling regulatory compliance and consumer communication. Therefore, packaging and sterilization remain indispensable in delivering safe and high-quality contact lenses to users worldwide. Continuous innovations in materials, technology, and processes are making these steps more efficient, effective, and sustainable, meeting both industry standards and environmental goals.

5.3 Recent Innovation: Integrating New Technologies into Lens Production

The contact lens industry has seen significant advancements over the years, driven by the need for improved comfort, safety, and visual performance. Recent innovations in manufacturing technologies are poised to revolutionize the production of contact lenses, making them more efficient, precise, and environmentally friendly. This section explores some of these cutting-edge technologies, with a special emphasis on ultra-violet light emitting diode (UV LED) curing, a promising process innovation that holds the potential to transform the industry [5].

One of the recent innovations is manufacturing contact lenses via 3D printing, as illustrated. The use of computer-aided design (CAD) software facilitates the creation of intricate prototypes that are then produced using 3D printing technology [29]. This advanced approach allows for the development of lenses with complex geometries, such as multifocal and toric designs, as well as therapeutic lenses intended for drug delivery. Moreover, rapid prototyping significantly shortens the development timelines for new materials and designs, promoting quicker innovation in the market. Despite these advancements, challenges such as achieving optimal surface smoothness and oxygen permeability of the lens persist, alongside the need for improved scalability in mass production and adherence to regulatory standards.

Another new technology in contact lens production is photopolymerization via UV LED. This technology is gaining traction in the contact lens industry due to its numerous advantages such as faster curing and energy-efficient compared to conventional UV curing systems [30]. The UV LED curing process involves exposing the contact lens material to UV light, which initiates a photochemical reaction that hardens the material. This method provides precise control over the curing process, resulting in lenses with superior optical clarity and mechanical properties.

Currently, UV LED photopolymerization is widely used to produce polymers. For instance, Ayub et al. [30] have developed a UV LED curing system for polyacrylamide (PAAm) and poly(N-isopropylacrylamide) (PNIPAAm) hydrogels. In this curing system, a new water-soluble photoinitiator (WSPI) was synthesized for effective photopolymerization. The hydrogels produced were transparent with 90% monomer conversion. Hence, the hydrogels produced via the UV LED curing approach have the potential for contact lenses and dental materials applications. Besides, another study was conducted by, where UV LED is used to cure polyurethane acrylate hydrophobic coating. The findings indicate that the coating has high C = C conversion (96–98%) and gel fraction (93–95%). This proved that UV LED photopolymerization can provide conversion and gel fraction of more than 90%, which is a promising technology to be used in producing polymer.

In UTM, our research group has collaborated with Alcon Ciba Vision Johor to develop a UV LED oven prototype for contact lens manufacturing. We came out with the idea to produce contact lenses via UV LED photopolymerization due to its energy efficient light source, with very specific and narrow emission output. The first stage of this development was to design the prototype oven and identify the suitable parameters such as the UV LED light intensity and curing time. Also, the energy required for the lens to be completely cured was estimated. This was achieved via a thermodynamic approach, where the rate of reaction for polymerization and the total polymerization time were determined by measuring the heat release using photo-differential scanning calorimetry (photo-DSC).

Our findings proved that UV LED can significantly reduce the curing time and cosmetic lens rejection compared to the traditional ultraviolet-mercury (UV-Hg) lamp. This development was further improved in the second stage by incorporating alternative photoinitiators in the contact lens formulation. This is because UV LED photopolymerization requires a photoinitiator with a maximum wavelength that overlaps the UV LED light emission. Therefore, the second stage study aims to ensure full polymerization via UV LED photopolymerization with a suitable photoinitiator to increase the output of the contact lenses.

The adoption of new technologies in contact lens manufacturing is not without challenges. One of the main obstacles is the high initial investment required for new equipment and training. Manufacturers need to weigh the long-term benefits of these technologies against the upfront costs. Additionally, regulatory approval processes can be lengthy and complex, delaying the introduction of new products to the market. To overcome these challenges, manufacturers can collaborate with technology providers and research institutions to share knowledge and resources. This can help accelerate the development and adoption of new technologies. Furthermore, demonstrating the cost-effectiveness and improved performance of these innovations can help gain regulatory approval and market acceptance.

The future of contact lens manufacturing looks promising, with continued advancements in technology driving improvements in product quality, efficiency, and sustainability. UV LED curing is just one example of how innovative processes can revolutionize industry. As more manufacturers adopt these technologies, we can expect to see a new generation of contact lenses that offer superior comfort, safety, and performance. In addition to UV LED curing, other emerging technologies such as 3D printing, smart materials, and biosensors will play a crucial role in shaping the future of contact lens manufacturing. By embracing these innovations, manufacturers can stay ahead of the competition and meet the evolving needs of consumers. Integrating new technologies into contact lens production is essential for improving the quality, safety, and sustainability of these products. UV LED curing stands out as a particularly promising innovation, offering numerous benefits over traditional curing methods. While there are challenges to adoption, the potential advantages make it a worthwhile investment for manufacturers. As the industry continues to evolve, embracing these advancements will be key to producing the next generation of contact lenses that meet the highest standards of performance and safety.

References

1. Jacobs, D.S., Carrasquillo, K.G., Cottrell, P.D., Fernández-Velázquez, F.J., Gil-Cazorla, R., Jalbert, I., Pucker, A.D., Riccobono, K., Robertson, D.M., Szczotka-Flynn, L., Speedwell, L., Stapleton, F.: BCLA CLEAR & #x2013; Medical use of contact lenses. Contact Lens Anterior Eye **44**, 289–329 (2021). https://doi.org/10.1016/j.clae.2021.02.002
2. Wallington, E.R.: Contact Lens Diameter: What is the Right Size?, https://delivercontacts.com/research/prescriptions/diameter/
3. Arroyo, R., Crespo, D., Alonso, J.: Influence of the base curve in the performance of customized and classical progressive lenses. Optom. Vis. Sci. **90**, 282–292 (2013). https://doi.org/10.1097/OPX.0b013e3182814d85
4. Tabernero, J., Klyce, S.D., Sarver, E.J., Artal, P.: Functional optical zone of the cornea. Invest. Ophthalmol. Vis. Sci. **48**, 1053–1060 (2007). https://doi.org/10.1167/iovs.06-0867

5. Shaker, L.M., Al-Amiery, A., Isahak, W.N.R.W.: Revolutionizing contact lens manufacturing: exploring cutting-edge techniques and innovations for enhanced vision and comfort. Int. J. Low-Carbon Technol. **19**, 359–385 (2024). https://doi.org/10.1093/ijlct/ctad136
6. Vincent, S.J., Alonso-caneiro, D., Collins, M.J.: Optical coherence tomography and scleral contact lenses: clinical and research applications. Clin. Exp. Optom. **102**, 224–241 (2019). https://doi.org/10.1111/cxo.12814
7. Bruce, I., Almond, S.L.: Demolding and delensing methods and systems in the manufacture of silicone hydrogel contact lenses, https://patents.google.com/patent/EP1752280A2/en#citedBy (2005)
8. Valint, P.L.J., Grobe, G.L.I., Ammon, D.M.J., Mcgee, J.A.: Plasma surface treatment of silicone hydrogel contact lenses with a flexible carbon coating, https://patents.google.com/patent/WO2000071611A1/en (1999)
9. Duskova-Smrckova, M., Podešva, J. í, Michálek, J. í: True story of poly(2-hydroxyethyl methacrylate)-based contact lenses: how did it really happen. Substantia. **6**, 79–91 (2022). https://doi.org/10.36253/Substantia-1591
10. Richard, G.B., David, D.W., Do-Hyung, K., David, B., Richard, F., Darren, R.: Novel automated process for aspheric surfaces. In: Proc.SPIE, pp. 445–450 (2000). https://doi.org/10.1117/12.405237
11. Banko, P.E.: Production of contact lenses by spin-casting. Clin. Exp. Optom. **59**, 286–289 (1976). https://doi.org/10.1111/j.1444-0938.1976.tb01447.x
12. Norbert, H., Christian, D., Thorben, B., Hendrik, R.: TiO_2 nanoparticles for enhancing the refractive index of hydrogels for ophthalmological applications. In: Proc.SPIE, p. 100780I (2017). https://doi.org/10.1117/12.2256296
13. Trzepieciński, T., Najm, S.M., Lemu, H.G.: Current concepts for cutting metal-based and polymer-based composite materials (2022). https://doi.org/10.3390/jcs6050150
14. Saptaji, K., Iza, N.R., Widianingrum, S., Mulia, V.K., Setiawan, I.: Poly (2-hydroxyethyl methacrylate) hydrogels for contact lens applications—a review (2021). https://doi.org/10.7454/mss.v25i3.1237
15. Abd Mana, N.A.M., Chin, L.C., Fook, C.Y., Yazid, H., Ali, Y.M.: A review on contact lens inspection. Indones. J. Electr. Eng. Comput. Sci. **31**, 700–712 (2023). https://doi.org/10.11591/ijeecs.v31.i2
16. Lee, J.H., Ha, B.H., Jang, D.K., Kim, Y.M., Jeong, Y.R., Han, S.M., Kim, K.H.: Development of optic coherence tomography device for contact lens inspection. J. Korean Ophthalmic Opt. Soc. **23**, 221–225 (2018). https://doi.org/10.14479/jkoos.2018.23.3.221
17. Moreddu, R., Vigolo, D., Yetisen, A.K.: Contact Lens Technology: From Fundamentals to Applications (2019). https://doi.org/10.1002/adhm.201900368
18. Coldrick, B.J., Richards, C., Sugden, K., Wolffsohn, J.S., Drew, T.E.: Developments in contact lens measurement: a comparative study of industry standard geometric inspection and optical coherence tomography. Contact Lens Anterior Eye. **39**, 270–276 (2016). https://doi.org/10.1016/j.clae.2016.01.002
19. Ellis, M.J., Noyce, T., Sluka, D.: Contact lens packages and methods of use, https://patents.google.com/patent/US20210394986A1/en?oq=US+20210394986A1 (2021)
20. Chetty, E.: Investigating the shelf life of disposable soft contact lenses. African Vis. Eye Heal. **67**, 89–94 (2008). https://doi.org/10.4102/aveh.v67i3.186
21. Complete Packaging Solutions: How the Heat Sealing Process for Blister Packs Work, https://completepackaginggroup.com/packaging-solutions/heat-sealing-blister-packaging/
22. Hamilton, R.S.: Packaging for contact lenses, https://patents.google.com/patent/US7225919B2/en?oq=US+7.225%2C919+B2 (2007)
23. IMA Automation: Processing, handling and packaging solutions, https://imautomation.com/en/contact-lens-solutions-technology-tcnl-13
24. Jildeh, Z.B., Wagner, P.H., Schöning, M.J.: Sterilization of objects, products, and packaging surfaces and their characterization in different fields of industry: the status in 2020. Phys. Status Solidi. **218**, 2000732 (2021). https://doi.org/10.1002/pssa.202000732

25. Lai, T.: An In-Depth Guide to Polypropylene—PP's Moisture-Resistant Properties in Packaging, https://patekpackaging.com/blogs/news/an-in-depth-guide-to-polypropylene-pps-moisture-resistant-properties-in-packaging
26. Drawell: Exploring Advanced Control Systems in Vertical Autoclaves: Enhancing Sterilization, https://www.drawellanalytical.com/exploring-advanced-control-systems-in-vertical-autoclaves-enhancing-sterilization/
27. Fink, J.K.: Contact lenses: chemicals, methods, and applications. John Wiley & Sons (2022)
28. Glasbey, T., Newman, S.D.: Methods and systems for contact lens sterilization, https://patents.google.com/patent/US8109064B2/en?oq=US+8%2C109%2C064+B2 (2012)
29. Alam, F., Salih, A.E., Elsherif, M., Yetisen, A.K., Butt, H.: 3D printed contact lenses for the management of color blindness. Addit. Manuf. **49**, 102464 (2022). https://doi.org/10.1016/j.addma.2021.102464
30. Ayub, N.F., Hashim, S., Jamaluddin, J., Adrus, N.: New UV LED curing approach for polyacrylamide and poly(: N -isopropylacrylamide) hydrogels. New J. Chem. **41**, 5613–5619 (2017). https://doi.org/10.1039/c7nj00176b

Chapter 6
Handling, Mishandling, and Care System of Contact Lenses

Abstract This Chapter offers valuable guidance for contact lens users, focusing on best practices for maintaining eye health and ensuring the safe use of lenses. It begins by encouraging users to consult optometrists or ophthalmologists, detailing the potential examinations and evaluations performed by eye care professionals. This chapter also provides practical, step-by-step instructions for applying contact lenses to the eye surface and emphasizes the importance of proper care and maintenance, including preventive measures to minimize mishandling. Additionally, the chapter addresses symptoms, risk factors, and complications associated with contact lens use, with a focus on issues arising from improper handling or the use of non-medical prescription lenses. It highlights the dangers of cheap, unregulated lenses and their potential to cause corneal damage, stressing the importance of choosing medical-grade products. Finally, the chapter explores challenges related to public perception of contact lens safety and emphasizes the need for user awareness. By equipping readers with the knowledge to detect symptoms, take preventive action, and make informed decisions, this chapter empowers users to safeguard their eye health and maximize the benefits of contact lenses.

Keywords Contact lens care · Lens handling · Hygiene protocols · Cleaning solutions · Complications · Professional care · Non-prescription lenses · User guidance · Maintenance · Safety

Contact lenses offer a convenient and aesthetically pleasing alternative to traditional eyeglasses, but their safe and effective use requires proper handling, care, and maintenance. Mishandling or neglecting care routines can lead to issues ranging from mild irritation to severe infections or complications that threaten vision quality. As regulated Class II or III medical devices, contact lenses are subject to strict safety standards, emphasizing the importance of hygiene, adherence to cleaning routines, and compliance with professional guidance to safeguard both lens performance and ocular health [1, 2].

This chapter provides a comprehensive guide to the handling, mishandling, and care system associated with contact lenses. It begins with foundational practices in

© The Author(s), under exclusive license to Springer Nature Singapore Pte Ltd. 2025

N. Adrus et al., *Contact Lenses*, Series in BioEngineering,
https://doi.org/10.1007/978-981-96-3636-5_6

Sect. 6.1 Getting Started with Contact Lenses, followed by detailed hygiene and care strategies in Sect. 6.2 Daily Care for Healthy Eyes and Lenses. Section 6.3 Recognizing and Addressing Risks focuses on identifying and mitigating complications, while Sect. 6.4 The Importance of Professional Care highlights the critical role of eye care professionals. By understanding and applying these guidelines, contact lens users can maintain their eye health, comply with safety regulations, and enhance their lens-wearing experience.

6.1 Getting Started with Contact Lenses

As a contact lens user, it is essential to understand that your lenses require a specialized prescription tailored to the unique shape and needs of your eyes. Unlike eyeglasses, contact lenses rest directly on the eye, which means your eye care professional will perform tests to measure corneal curvature, assess tear production, and ensure your eyes are healthy enough for lens wear. Equally important is learning how to insert and remove your lenses correctly. This includes maintaining proper hygiene, checking that your lens is oriented correctly, and carefully placing or removing it to avoid discomfort or damage. By understanding these basics, you can enjoy the benefits of contact lenses while ensuring your eye health and comfort.

6.1.1 Understanding Your Prescription: From Eye Exam to Lens Fitting

A comprehensive eye exam is a vital step in maintaining optimal eye health and overall well-being. Unlike basic vision screenings, which assess surface-level vision clarity, comprehensive exams delve deeper to uncover potential vision and eye health issues that may not exhibit obvious symptoms. Early detection and treatment of these issues are crucial for preventing vision loss and addressing underlying health conditions. During these evaluations, optometrists examine various aspects of vision and eye health to ensure precise diagnosis and effective treatment planning.

Key components of a comprehensive eye exam include gathering patient history to understand current eye or vision problems, overall health, medications, and environmental factors that may impact vision. Visual acuity tests, often using a Snellen chart, measure how clearly each eye can see at varying distances, providing a baseline for corrective needs (Fig. 6.1). Preliminary tests assess depth perception, color vision, eye muscle movements, peripheral vision, and pupil responses, which are essential for detecting subtle issues affecting visual performance.

Specialized tests, such as keratometry and corneal topography, measure corneal curvature to determine proper contact lens fit, optimizing both comfort and effectiveness. Refraction assessments pinpoint lens power for correcting refractive errors like

Fig. 6.1 Vision test using a Snellen eye chart (designed by Freepik, with slight modifications)

nearsightedness, farsightedness, or astigmatism, while evaluations of eye focusing, teaming, and movement ensure the eyes work harmoniously under varying visual demands. Advanced diagnostic instruments are also used to examine the internal and external structures of the eye, often with dilating drops for a detailed view, aiding in the early detection of serious conditions like glaucoma, diabetic retinopathy, and macular degeneration. Timely identification of such conditions is crucial for effective treatment and long-term vision preservation.

Eye exams are not limited to immediate diagnosis but also emphasize long-term vision care. Regularly scheduled exams, tailored to individual age and risk factors, play a pivotal role in catching problems early and managing them effectively. Figure 6.2 outlines the recommended frequency for comprehensive eye exams, which varies across age groups and risk categories [3]. For instance, children aged 6–17 should undergo an annual exam, while adults aged 18–39 are advised to have exams every two years. Older adults or those with risk factors, such as diabetes or family histories of eye disease, may require more frequent monitoring as advised by their eye care professional.

Baseline eye exams are particularly important for adults. The American Academy of Ophthalmology recommends that adults with healthy eyes and good vision have a complete eye exam once in their 20 s and twice in their 30 s [4]. This proactive approach helps detect potential issues early, ensuring they do not escalate into more serious problems. It also provides an opportunity to monitor changes in vision that may affect the suitability of contact lenses or other corrective measures. Proactive scheduling reflects the dynamic changes in vision that occur with growth, aging, and exposure to risk factors. For adults with no known conditions, baseline exams in their 20 s, followed by periodic checkups in their 30 s, can identify potential

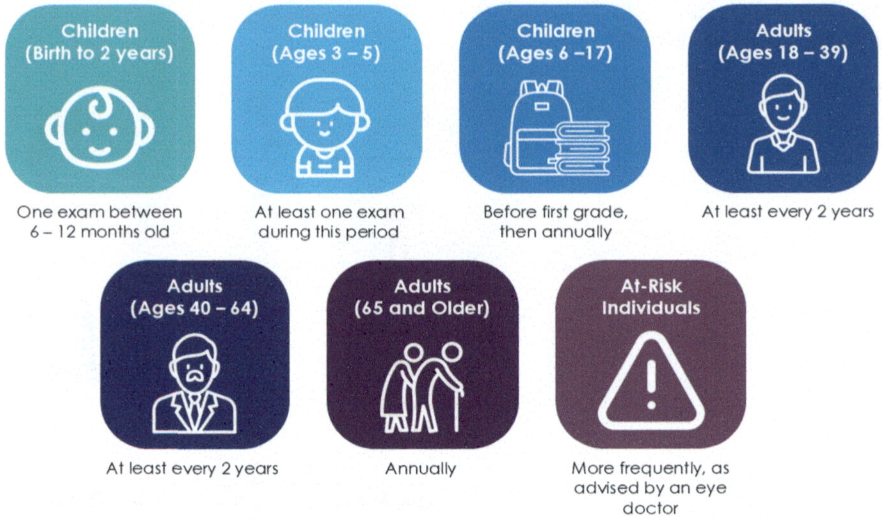

Fig. 6.2 Recommended frequency of eye examinations (adapted based on information from the *comprehensive pediatric eye and vision examination and comprehensive adult eye and vision examination, second edition* [3])

problems early, including issues that may affect the suitability of contact lenses. For high-risk individuals, such as those with diabetes or significant screen time, regular exams address conditions like digital eye strain, dry eyes, or corneal damage. Tailored recommendations ensure early management of these challenges and reduce the likelihood of complications [3].

By following recommended schedules and consulting their optometrists for personalized advice, individuals can safeguard their vision and overall eye health. These comprehensive evaluations provide accurate prescriptions while identifying potential complications, ensuring long-term ocular wellness. Regular exams not only support the effective use of contact lenses but also reduce the likelihood of serious complications. In prescribing and managing contact lenses for unique visual conditions, such as monocular visual impairment, comprehensive steps are required. These include accurate prescriptions [5], proper lens fitting, patient education, and follow-up care. By adhering to these methods, individuals can enjoy both improved vision and overall eye health, emphasizing the importance of comprehensive eye care.

Core Steps in Contact Lens Prescription and Management

Step 1: Detailed Eye Exam and Initial Assessment. Figure 6.3 provides a visual representation of the core steps in prescribing and managing contact lenses, ensuring an accurate and patient-centered approach to maintaining long-term vision health. The process begins with a thorough eye exam and initial assessment, which includes a detailed review of the patient's medical and ocular history, including prior eye conditions or surgeries. A comprehensive ocular examination, coupled with refraction tests, determines the patient's visual requirements and overall eye health. A dominant eye assessment is also conducted to analyze how the eyes function together, ensuring an accurate prescription and identifying potential serious ocular conditions.

In the dominant eye assessment, illustrated in Fig. 6.4, the individual focuses on an object while alternately covering each eye. The eye that remains aligned with the object is identified as the dominant eye. This evaluation provides valuable insights into vision clarity and depth perception, contributing to precise prescriptions and a better understanding of visual coordination.

Step 2: Prescription Parameters. Prescription parameters are crucial for ensuring lenses provide optimal vision correction and comfort. Thebase curve (BC) determines the curvature of the lens for a proper corneal fit, while the diameter (DIA) specifies the lens size to adequately cover the cornea. Power/sphere (PWR) adjusts for refractive errors like myopia or hyperopia, and additional parameters such as cylinder (CYL) and axis are included for astigmatism correction. Multifocal adjustments or partial corrections may also be required to balance vision quality in specific cases. These parameters ensure the lenses align perfectly with the patient's unique ocular needs.

Step 3: Lens Fitting. Lens fitting is carried out to verify proper alignment, movement, and comfort. Trial lenses, often made of high oxygen-permeable materials like silicone hydrogel, are fitted to minimize the risk of corneal hypoxia. For specialized corrections, such as astigmatism or keratoconus, Rigid gas-permeable (RGP) or

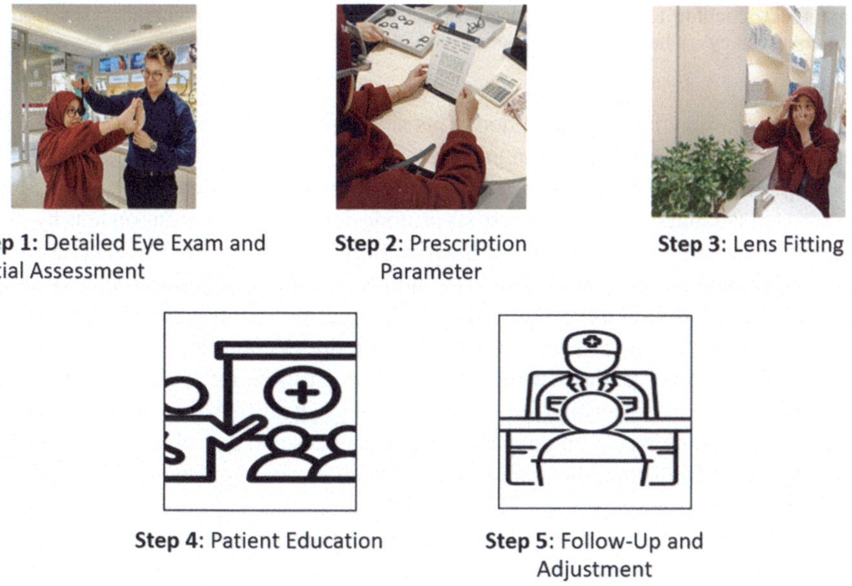

Fig. 6.3 Flowchart of comprehensive steps in prescribing and managing contact lenses

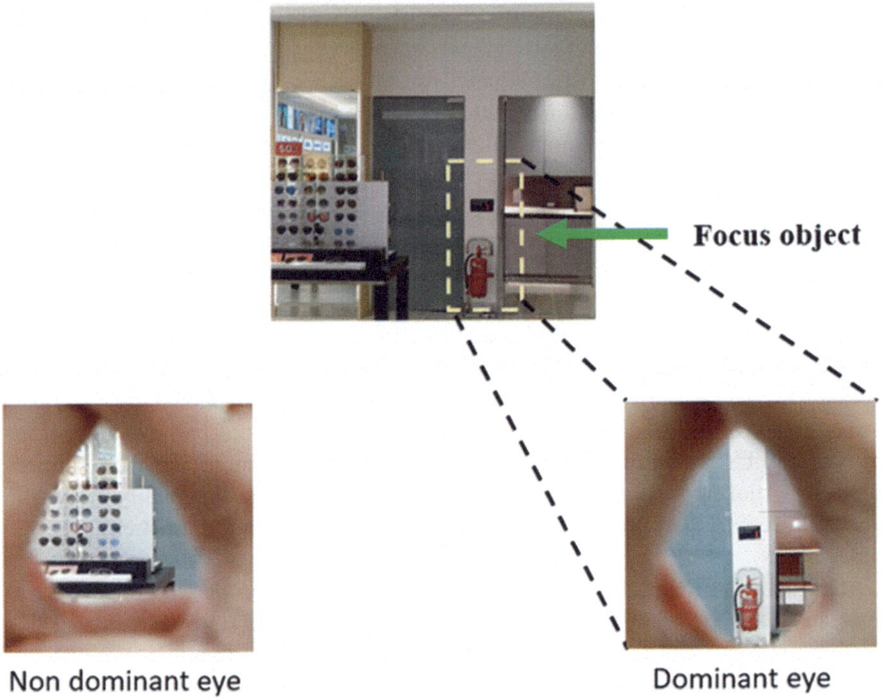

Fig. 6.4 Image of the dominant eye test. The focus object (a fire extinguisher) is viewed through a triangular opening formed by the hands. The image on the left represents the view from the non-dominant eye, where the object appears shifted

scleral lenses may be recommended. This step ensures the lenses accommodate the patient's eye anatomy and vision goals.

Step 4: Patient Education. Patient education is a cornerstone of successful contact lens management. Optometrists provide detailed instructions on lens hygiene, wearing schedules, and proper handling to minimize risks like microbial keratitis. For monocular patients, the discussion also emphasizes the added importance of proper care, as these individuals rely heavily on one eye for vision. Empowering patients with knowledge ensures healthy practices and reduces complications.

Step 5: Follow-Up and Adjustments. Regular follow-up appointments monitor the performance of prescribed lenses and overall eye health. Adjustments to prescriptions may be made based on the patient's experiences and any observed issues. This ongoing care reinforces the importance of both short-term vision correction and long-term ocular wellness.

By incorporating these steps, the process of prescribing and managing contact lenses ensures that patients receive accurate prescriptions, comfortable fittings, and the knowledge needed to maintain their vision health over time. This holistic approach highlights the importance of regular eye care in addressing both immediate and future vision needs.

Following these comprehensive steps in prescribing and managing contact lenses, it is important to understand the fundamental differences between eyeglasses and contact lens prescriptions. Contact lens and eyeglass prescriptions differ fundamentally due to the distinct ways each lens interacts with the eye. Eyeglasses are positioned approximately 12 mm away from the eyes, while contact lenses rest directly on the corneal surface [6]. This proximity affects the corrective power required, necessitating separate prescriptions for each. Contact lens prescriptions are specifically designed to account for their placement on the cornea, ensuring accurate vision correction and comfortable wear. As summarized in Table 6.1, these differences also extend to the requirements of eye exams, with contact lenses needing additional evaluations to ensure proper fit and long-term eye health.

Separate eye exams are necessary for contact lenses and eyeglasses due to their differing requirements. A comprehensive eye exam for eyeglasses focuses on assessing overall eye health and determining lens power to correct refractive errors. In contrast, a contact lens exam includes additional evaluations specific to fitting lenses on the cornea: (i)corneal measurements to assess curvature and size for proper lens fit, (ii) tear film evaluation to ensure sufficient tear production for comfortable wear, and (iii) trial lens fitting to test lenses on the eye and evaluate vision and comfort before finalizing the prescription. As summarized in Table 6.1, contact lens exams require additional measurements and procedures that are unnecessary for eyeglasses. These assessments are crucial to ensuring that contact lenses not only correct vision effectively but also fit securely and promote long-term ocular health.

Contact lens prescriptions also include parameters not found in eyeglass prescriptions. These specifications, such as the BC and DIA, ensure the lenses align with

Table 6.1 Differences between eye exams for glasses and contact lenses

Aspect	Glasses eye exam	Contact lens eye exam
Lens placement	Lenses sit approximately 12 mm from the eyes	Lenses rest directly on the surface of the eyes
Prescription	Includes power for vision correction	Includes power, base curve (BC), diameter (DIA), and lens material or brand
Additional measurements	Not required for glasses	Includes corneal curvature, size, and tear film evaluation
Tear film evaluation	Not assessed	Evaluates tear production to ensure sufficient moisture for comfortable lens wear
Trial lenses	Not applicable	Trial lenses are fitted to test comfort
purpose	Corrects refractive errors for glasses	Ensures proper fit, safety, and suitability of contact lenses
Health considerations	Focuses on overall eye health and vision correction	Includes additional assessments specific to contact lens safety and comfort

the natural shape of the cornea for a secure and comfortable fit. Additionally, the lens material or brand must be specified to account for oxygen permeability, which directly affects comfort and eye health. For individuals with astigmatism, parameters such as cylinder (CYL) and axis are added to correct irregularities in corneal curvature. Multifocal lens prescriptions may also include addition (ADD) for magnifying power and dominant (D/N) indicators for eye dominance. These tailored measurements ensure that contact lenses provide precise optical correction and promote wearer comfort.

As summarized in Table 6.1, eyeglass prescriptions do not include these parameters, making them unsuitable for purchasing contact lenses. Using an eyeglass prescription to order contact lenses can result in poor fit, discomfort, and even complications such as corneal irritation or reduced oxygen supply. This underscores the importance of obtaining a contact lens-specific exam and prescription tailored to the unique needs of the wearer.

To further demonstrate how these parameters are applied in practice, Table 6.2 presents an example prescription for monthly MultiFocal Soflens, which includes all the essential specifications for proper fit and functionality. This prescription highlights the precision required to align the lens with the corneal curvature through measurements such as base curve and diameter. Additionally, cylinder and axis values address astigmatism, while addition and dominant parameters ensure customization for multifocal lenses to correct presbyopia. These specifications emphasize the attention to detail involved in prescribing contact lenses.

A complete contact lens prescription also includes the lens brand (e.g., AirOptix) and an expiry date (Fig. 6.5). These details guarantee the lenses are used within the recommended duration and conform to the manufacturer's guidelines, minimizing risks of complications. Using expired lenses can lead to discomfort, reduced vision quality, and increased risk of infections. Adhering to these prescription guidelines ensures optimal vision correction and long-term ocular health.

Ensuring proper fit is critical for contact lenses to provide clear vision and maintain ocular health. Unlike eyeglasses, which are positioned at a distance from the eyes, contact lenses rest directly on the corneal surface. An improper fit can lead to discomfort, irritation, or serious complications such as corneal abrasions and infections. Accurate measurements and regular follow-ups with an eye care professional help mitigate these risks and ensure lenses meet the unique anatomical and optical needs of the wearer.

The risks of using an eyeglass prescription for contact lenses further emphasize the importance of accurate fitting and professional guidance. Eyeglass prescriptions lack the corneal measurements, tear film evaluations, and lens-specific parameters

Table 6.2 Example contact lens prescription of one of the authors

Eye	Base curve (mm)	Diameter (mm)	Power/Sphere	Cylinder	Axis	Addition
Left	8.8	14.5	−2.25 D	−0.75 D	110	+2.50
Right	8.8	14.5	−3.50 D	−0.50 D	75	+2.50

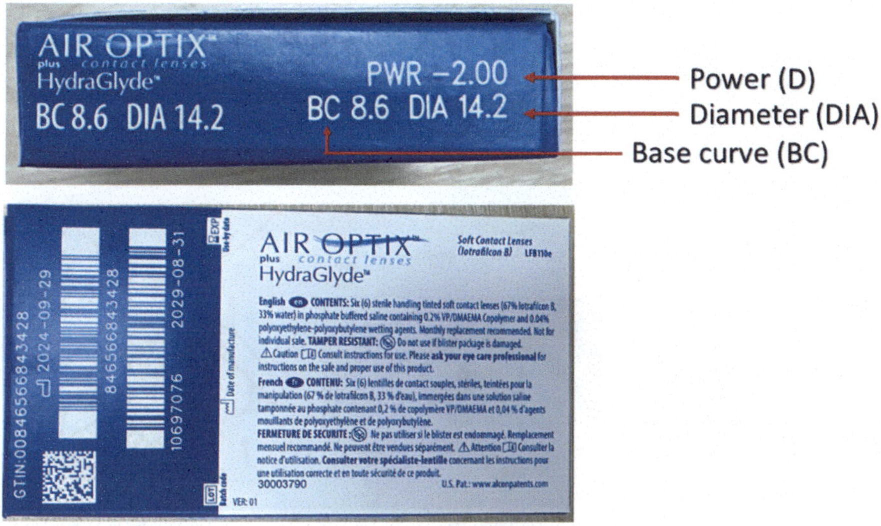

Fig. 6.5 Example contact lens packaging details, courtesy from Alcon Johor

required for safe and effective contact lens wear. This can result in lenses that do not fit securely or provide sufficient oxygen to the cornea, increasing the likelihood of irritation, infection, and long-term complications.

By understanding the differences in prescriptions and adhering to contact lens-specific recommendations, wearers can ensure optimal vision correction, comfort, and long-term eye health. Regular consultations with an optometrist safeguard against potential risks and provide the necessary adjustments to maintain healthy and effective lens use over time.

6.1.2 Mastering the Basics: A Step-By-Step Guide to Insertion and Removal

The insertion of contact lenses requires meticulous preparation, adherence to hygiene standards, and precise technique [7]. As shown in Fig. 6.6, the process begins with assembling essential items in a well-lit room, including contact lenses, a suitable solution, a mirror, and clean materials for drying hands. Hands should be thoroughly washed with plain, non-moisturizing soap to remove any residue or contaminants (Step 1). After rinsing, hands must be dried using a clean, lint-free towel or disposable paper towels to prevent contamination. Keeping fingernails trimmed is also recommended to avoid scratching the lens or injuring the eye [8].

The lens is carefully removed from its case or blister pack, placed in the palm of the hand, and wetted with a storage solution (Step 2). Even if a no-rub solution is used, gently rubbing the lens with a fingertip ensures thorough cleaning. Before

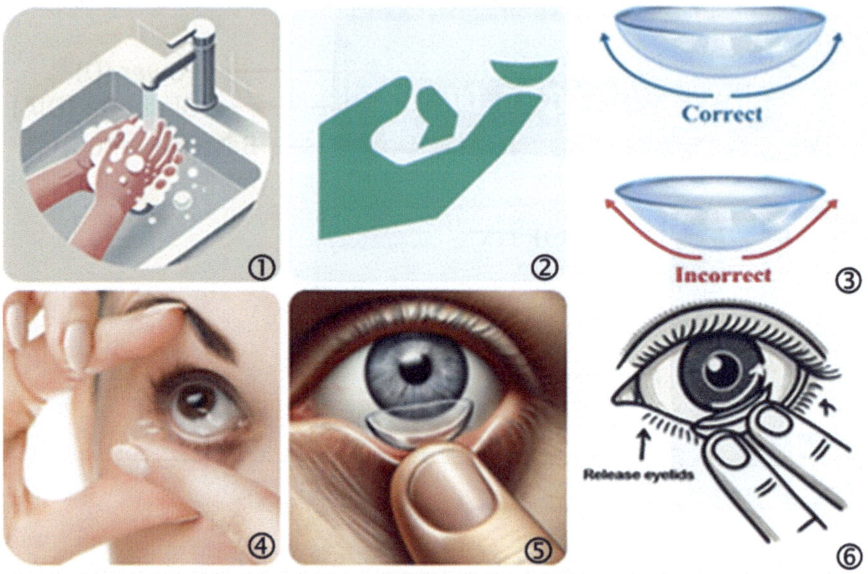

Fig. 6.6 Step-by-step guide for contact lens insertion, illustrating seven essential steps: (1) Preparation—wash hands thoroughly, (2) Lens Handling—clean the lens with solution, (3) Inspect the Lens—ensure proper orientation, (4) Hold Eyelids Open—use fingers to open upper and lower eyelids, (5) Place the Lens—gently place it on the cornea or lower white part of the eye, (6) Release Eyelids—close the eye and press lightly to remove air bubbles, and (7) Adjust for Comfort—remove, clean, and reinsert the lens if discomfort occurs. Repeat the process for the second eye

insertion, the lens must be balanced on the tip of the index finger and inspected for proper orientation (Step 3). A correctly oriented lens has a smooth, bowl-like shape, whereas an inside-out lens has edges that flare outward. Proper positioning is essential for both comfort and function [9].

To insert the lens, the non-dominant hand holds the upper eyelid open, while the middle finger of the dominant hand pulls the lower eyelid downward (Step 4). Looking straight ahead or slightly upward, the lens is gently placed onto the cornea or the lower white part of the eye (Step 5). After placement, the eyelids are slowly released, and the eye is briefly closed. Light pressure is applied to the upper eyelid to remove air bubbles and ensure the lens settles correctly (Step 6). If any discomfort occurs, the lens should be removed, cleaned with the solution, and reinserted to avoid irritation (Step 7). The process is then repeated for the other eye [7, 9].

The clear, step-by-step process depicted in Fig. 6.6 simplifies the contact lens insertion procedure, emphasizing the importance of hygiene, correct lens orientation, and comfort adjustments. Following these steps carefully ensures a safe and comfortable wearing experience.

Removing contact lenses is just as essential as their insertion, requiring careful hygiene and precise technique to maintain eye health and ensure the lenses remain clean and intact. Begin by washing hands thoroughly with plain, non-moisturizing soap and drying them completely with a clean, lint-free towel. This ensures that dirt or bacteria are removed and that fingers can handle the lens effectively. Always work over a clean, well-lit surface, such as a sink with the drain plugged, to prevent the lens from being lost if dropped.

Start with the right eye for consistency and prepare the lens case by filling it with fresh contact lens solution. Confirm that the lens is properly centered on the cornea, applying rewetting drops if the eye feels dry to ease removal. Look upward and use the middle finger of the dominant hand to gently pull down the lower eyelid. With the index finger, slide the lens downward to the lower white part of the eye, then gently pinch the lens between the thumb and index finger to lift it off the eye. Apply minimal pressure to avoid damaging the lens or causing discomfort. Repeat this procedure for the left eye.

After removal, clean the lens thoroughly using a fresh multipurpose solution, following the manufacturer's instructions for proper disinfection. Place the lens in its respective compartment in the lens case, ensuring it is fully submerged in the solution. For extended-wear lenses, always use cases labeled "L" and "R" to prevent mixing. For daily disposable lenses, discard them immediately after removal. These steps, as depicted in Fig. 6.7, illustrate a clear and methodical process to ensure safe removal and proper care, minimizing the risk of eye irritation or infection.

6.2 Daily Care for Healthy Eyes and Lenses

6.2.1 Choosing the Right Care System: Understanding Different Types and Schedules

Selecting the appropriate contact lens care system is essential for maintainingeye health and preventing complications like infections or irritation. The choice of a care system depends on several factors, including lens type, material, lifestyle, and individual eye health needs. Consulting an eye care professional ensures that the selected system aligns with the user's requirements and safety guidelines [3]. **Multipurpose solutions** are a widely used option due to their convenience, as they combine cleaning, disinfecting, rinsing, and storage in one product. These solutions are ideal for users seeking a streamlined routine without compromising hygiene. For individuals with allergies or sensitive eyes, **hydrogen peroxide-based systems** provide an alternative with superior disinfection capabilities. These systems use specialized cases to neutralize the solution and must be handled with care to avoid irritation from unneutralized peroxide. Lastly, **saline solutions** are primarily used for rinsing lenses but lack disinfecting properties and must be paired with other care products to maintain sterility.

Other care system should also consider is the type of lenses used. For example, daily disposable lenses, are single-use and eliminate the need for cleaning or disinfection, making them highly convenient for individuals with busy routines or sensitive

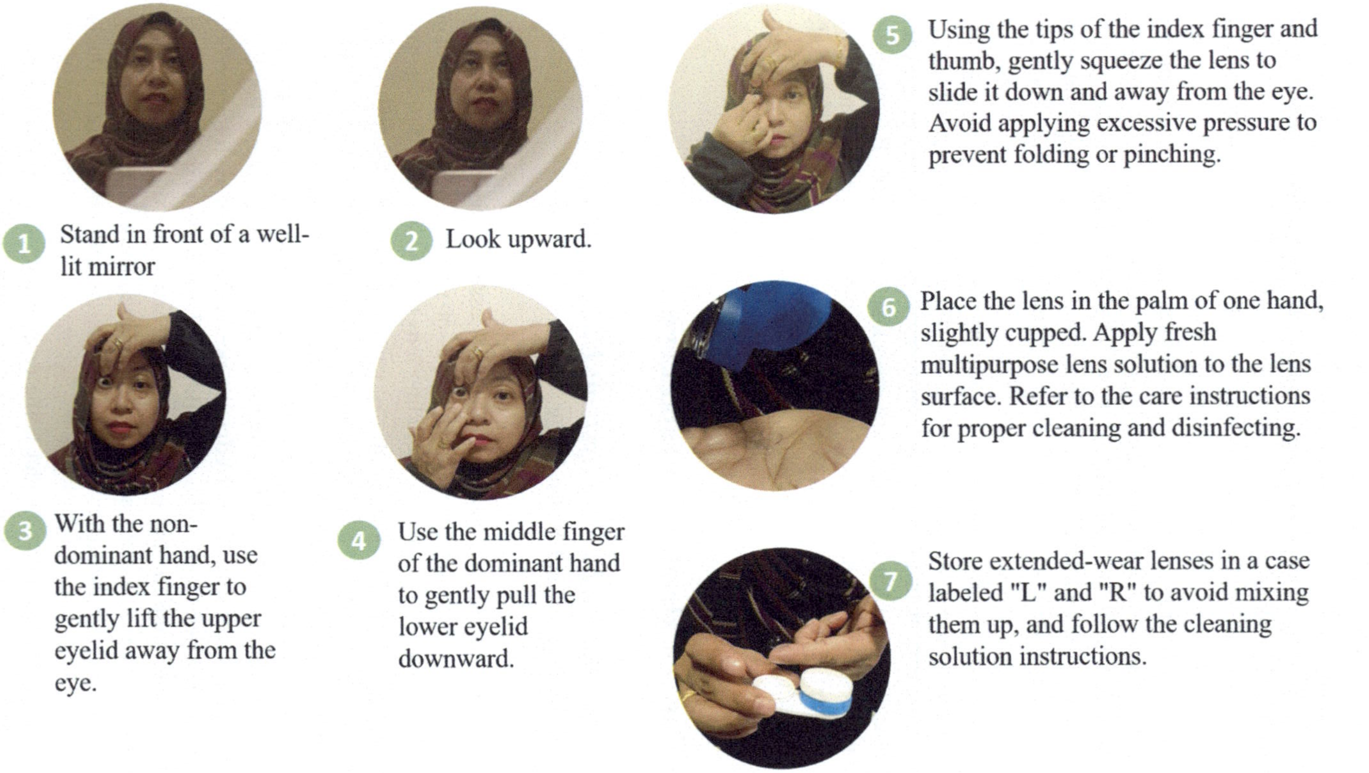

Fig. 6.7 Step-by-step guide for contact lens removal: (1) Prepare by washing and drying hands and stand in front of the mirror, (2) Look upward, (3) Gently lift the upper eyelid using the non-dominant hand, (4) Pull the lower eyelid downward using the middle finger, (5) Slide the lens downward with the index finger, (6) Pinch the lens gently to lift it off the eye, (7) Store the lenses in a labeled case with fresh solution or discard if using daily disposables

Table 6.3 Do's and Don'ts in contact lens care

Do's	Don'ts
Use fresh solution for every cleaning cycle	Do not use tap water, saliva, or homemade saline
Rinse lenses thoroughly after rubbing	Avoid topping off old solution in the lens case
Follow the manufacturer's recommended storage time	Do not switch care products without consulting an eye care professional
Wait for peroxide neutralization before wearing	Never rinse lenses with unnaturalized hydrogen peroxide
Consult your optometrist regularly	Avoid wearing lenses in water environments like pools or bathtubs

eyes. Conversely, **reusable lenses**, such as monthly or biweekly lenses, require more comprehensive care, including regular cleaning and disinfecting. Some lens materials, like **silicone hydrogel**, may benefit from specific cleaning solutions designed to prevent residue buildup that could affect comfort or clarity.

Lifestyle also considered as further influence the choice of a care system. For example, frequent travelers or individuals with demanding schedules might prefer compact multipurpose solutions that simplify the cleaning and storage process. For those participating in activities such as swimming or sports, water can harbor microorganisms like *Acanthamoeba*, which can cause serious eye infections. By selecting the right care system and lenses. Regular consultations with an eye care professional are crucial for tailored advice, especially when switching lens types or addressing specific eye health concerns. Adhering to the recommended care system ensures optimal lens performance, reduces the risk of complications, and promotes long-term eye health. Table 6.3 summarizes key safety practices and essential guidelines for effective lens care.

In conclusion, selecting the right contact lens care system is essential for maintaining eye health and preventing infections. By understanding the types of solutions available and practicing proper hygiene, such as using fresh solution, avoiding exposure to water, and cleaning lenses regularly, wearers can significantly reduce the risk of complications. Regular consultations with an eye care professional are equally important to ensure the chosen care system remains suitable and to monitor eye health. Adhering to these safety guidelines allows contact lens users to enjoy safe, comfortable, and long-lasting lens wear.

6.2.2 Cleaning and Storing Your Lenses: Essential Hygiene Practices

Proper cleaning and storage of contact lenses are essential for maintaining eye health and ensuring the longevity of your lenses. Before handling lenses, always wash and dry your hands thoroughly using soap and water. Even seemingly clean hands can carry invisible oils, dirt, or bacteria that may transfer to the lenses, causing irritation or infections. Dry your hands with a lint-free towel to avoid contaminating the lenses. Once your hands are clean, follow the recommended cleaning steps: gently rub the lenses with your fingers to remove debris, then rinse them with fresh contact lens solution. After cleaning, soak the lenses overnight in the recommended solution to disinfect them and keep them in optimal condition [10]. Figure 6.8 highlights common mistakes, such as neglecting hand hygiene or sleeping with lenses, which can lead to serious infections.

When it comes to storing your contact lenses, always have a clean lens case and ensure the lenses are fully submerged in fresh solution. Never reuse old solution, as it loses its disinfecting properties and can lead to bacterial contamination. Clean your lens case daily by rinsing it with hot water or contact lens solution and allowing

Fig. 6.8 Infographic of common contact lenses mistakes

it to air dry. Replace the lens case every three months or sooner if it shows signs of wear, such as cracks or discoloration. These practices significantly reduce the risk of infections and discomfort while extending the life of your lenses [3].

If makeup or debris gets stuck on the lenses, avoid using water or saliva for cleaning, as these can introduce harmful microorganisms. Instead, use contact lens solution to gently remove particles. After cleaning, thoroughly rinse the lenses with fresh solution before storing them. For lenses that are not worn daily, clean and disinfect them before reuse, even if they have been stored in solution for an extended period. This ensures the lenses remain free of bacteria and other harmful contaminants that could cause irritation or infection [11].

Adhering to the prescribed replacement schedule for your lenses is equally critical for preventing complications. Overwearing lenses beyond their recommended duration can result in discomfort, blurry vision, and a higher risk of eye infections. Regular eye exams are essential to monitor your eye health, ensure your prescription is up-to-date, and identify any potential issues early. Mishandling and improper care of lenses can increase the risk of contamination and lead to serious infections. These mishandling issues are explored further in this Chapter, Sect. 6.3.1. By following a consistent cleaning routine, storing lenses correctly, and consulting your eye doctor regularly, you can minimize the risk of complications and maintain safe, comfortable, and effective use of your contact lenses.

6.2.3 Wearing Time: Guidelines for Daily and Extended Wear

Contact lenses are designed with specific wearing schedules to balance comfort and eye health. **Daily wear contact lenses** are intended for use only during waking hours and must be removed before sleeping to prevent complications such as reduced oxygen supply to the cornea. This limitation minimizes the risk of irritation, infections, and serious conditions like keratitis. In contrast, **extended-wear lenses** are made from advanced materials like highly breathable silicone hydrogel or RGP materials, which allow higher oxygen permeability. These lenses can be worn continuously for up to seven to thirty days, depending on the lens type and the recommendations of an eye care professional. However, extended-wear lenses carry a higher risk of complications, such as corneal ulcers and infections, especially if hygiene practices are not strictly followed [12].

The choice between daily and extended-wear lenses depends on individual preferences, lifestyle, and health considerations. **Daily disposable lenses** are ideal for individuals seeking convenience and hygiene. Their single-use nature eliminates the need for cleaning or storage, significantly reducing the risk of bacterial buildup. They are particularly suitable for users with sensitive eyes or busy routines, as they offer a fresh, sterile pair of lenses each day. On the other hand, extended-wear lenses provide the advantage of continuous vision, making them convenient for users who prefer not to handle lenses daily [13]. These lenses are especially appealing to individuals with demanding schedules or frequent travel needs. However, they require careful cleaning and disinfecting whenever removed, and improper use such as wearing them beyond the recommended period or exposing them to water during activities like swimming can greatly increase the risk of infections [14].

Table 6.4 summarizes the key differences in hygiene practices between daily and extended-wear lenses. Daily lenses are simple, requiring no cleaning or storage, and are disposed of after a single use, making them low-risk and time-efficient. Extended-wear lenses, on the other hand, demand strict adherence to cleaning and storage protocols to avoid contamination. These lenses must be stored in a clean case with fresh disinfecting solution, and the case should be replaced every three months to maintain hygiene.

A randomized clinical trial involving 50 participants compared the performance of daily disposable and extended-wear lenses in terms of visual acuity, comfort, and ocular health. Both lens types showed similar results, with comparable wearing times and user satisfaction. Interestingly, many participants preferred extended-wear lenses due to the convenience of avoiding daily changes. This highlights the importance of aligning lens type with individual habits and preferences. While daily lenses offer superior hygiene with minimal effort, extended-wear lenses are practical for those willing to adhere to rigorous cleaning routines [15].

When choosing between daily and extended-wear lenses, factors such as cost, lifestyle, and commitment to proper care should be carefully considered. Daily lenses, although more expensive upfront, provide unmatched convenience and reduce the risk of complications due to improper cleaning or storage. Conversely, extended-wear lenses are more cost-effective over time but require a higher level of maintenance to ensure safety. Consulting an eye care professional is essential to determine the most suitable option based on individual needs, eye health, and willingness to follow care guidelines. Both lens types are viable choices, but success ultimately depends on the wearer's preferences and adherence to recommended practices.

Table 6.4 Comparison of hygiene practices for daily and extended-wear contact lenses

Aspect	Daily contact lenses	Extended wear contact lenses
Cleaning requirements	No cleaning needed; dispose of after single use	Requires cleaning and disinfecting after removal
Handling	Wash hands before inserting lenses to avoid contamination	Wash hands before handling and ensure proper cleaning
Storage	Not required; disposed of after use	Store in a clean case with fresh disinfecting solution
Replacement of case	Not applicable	Replace the storage case every three months
Risk of contamination	Minimal due to single-use nature	Higher without consistent cleaning and proper storage
Routine adherence	Simple and less time-consuming	Demands strict adherence to cleaning and storage practices

6.3 Recognizing and Addressing Risks

6.3.1 Common Mishandling Mistakes in Contact Lens Handling

Improper handling of contact lenses is a significant contributor to eye health complications, including infections, discomfort, and serious corneal damage. One of the most common mistakes isoverwearing contact lenses, as wearing them beyond the recommended duration disrupts oxygen supply to the cornea. The cornea, which lacks blood vessels, relies on oxygen absorbed through the tear film to maintain its health and function [16]. Contact lenses act as a partial barrier to oxygen transmission, as lenses and as they age or are worn for extended periods, they increasingly obstruct this flow. This disruption can lead to conditions such as **corneal hypoxia**, thinning, and edema [17]. Even high-oxygen-permeable lenses, such as certain types of silicone hydrogel lenses approved for extended wear, can pose risks if worn continuously without following the prescribed schedule. Using non-extended-wear lenses overnight or beyond recommended durations significantly increases the risk of epithelial thinning, slower cell renewal, and infiltrative complications like corneal ulcers [18].

Beyond wearing habits, poor hygiene during contact lens handling is another major contributor to complications. Failing to wash hands before handling lenses transfers bacteria, oils, and dirt to the lens surface, increasing the risk of microbial keratitis and other infections. Studies reveal widespread contamination, with up to 34% of lens cases, 9% of lenses, and 11% of cleaning solutions found to harbor harmful microorganisms [19, 20]. Consistent hygiene practices, such as washing hands thoroughly with antibacterial soap, drying them with a lint-free towel, and cleaning lens cases daily, are crucial. Manual rubbing of lenses, even when using "no-rub" solutions, is also essential to remove debris and microorganisms effectively. Peroxide-based solutions are recommended for individuals sensitive to preservatives, particularly those using RGP lenses.

Delaying lens replacement is another common mistake. Research shows that nearly half of disposable and frequent-replacement lens users wear their lenses beyond the recommended timeframe, often waiting until discomfort occurs to replace them [21]. This behavior significantly increases the accumulation of protein deposits, debris, and bacteria, compromising lens integrity and raising infection risks. Disposable lenses should be discarded after each use, while other lenses must adhere to replacement intervals advised by manufacturers or eye care providers. Using expired lenses or stretching replacement schedules increases the risk of serious infections, such as microbial keratitis, which can cause symptoms ranging from irritation and redness to severe pain and vision loss [22]. Timely intervention and adherence to replacement schedules are essential to prevent such outcomes.

Sleeping in contact lenses is another high-risk behavior that significantly exacerbates complications. During sleep, closed eyelids restrict oxygen supply to the cornea, creating ahypoxic environment that exacerbates complications like corneal neovascularization, inflammation, and infections. While extended-wear lenses made from silicone hydrogel provide enhanced oxygen permeability, they cannot eliminate these risks. Approved extended-wear lenses should still be removed periodically to allow the eyes to recover and maintain corneal health [13].

Exposure to water is another critical concern for contact lens wearers. Activities like swimming, showering, or using hot tubs introduce harmful microorganisms, such as *Acanthamoeba*, which may lead to severe infections such as *Acanthamoeba keratitis*. Water exposure also traps bacteria against the lens surface, increasing the risk of contamination and corneal damage. Unless specifically designed for water exposure, contact lenses should always be removed before engaging in these activities to safeguard eye health [23].

Lastly, lifestyle factors and external conditions also play a critical role in lens safety. Smoking is a notable risk factor, as it degrades the lipid layer of the tear film, leading to dryness and irritation, while the chemicals in tobacco smoke reduce corneal sensitivity. Additionally, medications such as diuretics, antihistamines, and psychotropic drugs interfere with tear production, exacerbating dryness and discomfort for lens users. Steroids and immunosuppressants weaken immune defenses, increasing susceptibility to infections when hygiene is neglected [21]. These factors highlight the importance of tailoring lens use and care to individual circumstances, including addressing environmental and systemic health conditions.

In conclusion, understanding and avoiding common mishandling mistakes such as overwearing lenses, neglecting hygiene, delaying replacements, sleeping in lenses, or exposing them to water can significantly reduce the risk of complications. Adhering to proper care practices, timely replacement schedules, and professional guidance ensures that contact lens wearers can enjoy safe, comfortable, and effective vision correction. Early recognition of symptoms like redness, discomfort, or blurry vision, followed by prompt intervention, is vital for protecting long-term ocular health.

6.3.2 Symptoms to Watch Out for: Early Detection for Prompt Action

The initial experience of wearing contact lenses can involve a variety of sensations as the eyes adapt to their presence. While most symptoms are temporary and part of the natural adjustment process, it is important to recognize those that may require professional attention. Understanding these symptoms and their causes ensures a smoother transition and helps maintain eye health. Table 6.5 summarizes common symptoms experienced by new contact lens wearers, their causes, possible solutions, and recommendations for seeking professional advice.

One of the most common sensations for new contact lens wearers is lens awareness, which may feel like slight movement or a general presence on the cornea. This is a normal and positive sign that the lens is functioning correctly. The natural flow of tears and blinking allows the lens to shift slightly, promoting proper hydration,

Table 6.5 Common symptoms experienced by new contact lens wearers

Symptom	Cause	Solution	When to seek professional advice
Lens awareness	Normal lens movement and adjustment on the cornea	Allow time for the eye to adjust; symptoms usually resolve within 15 min	If discomfort persists or is accompanied by pain, dryness, or redness
Excessive tearing	Eye's natural response to a foreign object	Wait for adaptation; use proper techniques for insertion and removal	If tearing continues beyond the initial period
Light sensitivity	Temporary adjustment to the lens	Wear sunglasses in bright environments	If sensitivity worsens or persists
Burning sensation	Dryness or inadequate lubrication	Use lubricating drops designed for contact lenses	If burning continues despite using drops or is severe
Blurry vision	Debris, oils, or tears on the lens surface	Remove, clean, and reinsert the lens	If blurriness persists despite cleaning or indicates a prescription issue
Off-center lens	Lens movement due to improper fit or dryness	Gently blink or massage eyelid; clean and reinsert if needed	If the lens repeatedly moves out of position
Excessive blinking	Natural response to lenses, promoting moisture distribution	Ensure lens hydration and fit	If blinking becomes disruptive or excessive over time
Red eyes	Dryness, irritation, or reaction to lens materials	Use saline solution or lubricating drops	If redness persists or is severe
Itchy eyes	Dryness, allergens, or improper lens fit	Use lubricating eye drops	If itching continues or is accompanied by redness or swelling

oxygen transmission, and tear exchange. These processes are essential for maintaining eye health [19]. Typically, this sensation subsides within 15 min, but if it persists, it may indicate dryness, improper fit, or the need for a different lens material. Consulting an eye care professional can help resolve these issues [19].

Excessive tearing is another frequent symptom caused by the eye's natural response to the lens as a foreign object. This tearing lubricates the lens and protects the eye, typically subsiding as the eyes adjust. Proper insertion techniques can minimize discomfort, but if tearing continues beyond the adjustment period, professional advice should be sought. Similarly, light sensitivity is a common adaptation symptom. Wearing sunglasses in bright environments can alleviate discomfort, but worsening or prolonged sensitivity warrants consultation with an eye care provider [24].

Next, burning sensations and dryness are also common issues. These symptoms often stem from inadequate lubrication and can be alleviated with lubricating eye drops designed for contact lenses. If burning persists despite using drops, it could indicate an allergic reaction, poor lens fit, or another underlying problem. Blurry vision, often due to debris or oils on the lens, can usually be resolved by removing, cleaning, and reinserting the lens. Persistent blurriness, however, may signal a prescription issue or improper fit requiring professional evaluation [24].

Off-center lenses can cause discomfort but are generally easy to fix with a gentle blink or light massage of the eyelid. If the lens repeatedly shifts out of place, dryness or a poor fit may be the cause, and an eye care provider should be consulted. Red and itchy eyes are common symptoms linked to dryness, irritation, or hygiene issues. Persistent redness or itching, especially when accompanied by swelling or discomfort, may indicate serious conditions such as Contact Lens-Induced Acute Red Eye (CLARE) or Giant Papillary Conjunctivitis [25]. Switching to hypoallergenic products or alternative lens materials may help, but professional advice is essential for proper treatment [24]. High-risk behaviors, including sleeping in lenses without approval, overwearing them, or exposing lenses to water during activities like swimming, increase the likelihood of complications.

Water exposure introduces microorganisms such as *Acanthamoeba*, which can cause severe infections like *Acanthamoeba keratitis*. Adhering to hygiene practices and avoiding these risks ensures safer lens use. Figure 6.9 illustrates the common causes of red eyes, including dryness, improper fit, extended wear, and hygiene-related issues.

Most of these symptoms subside as the eyes adjust to wearing contact lenses. However, persistent or worsening issues like redness, burning, blurry vision, or

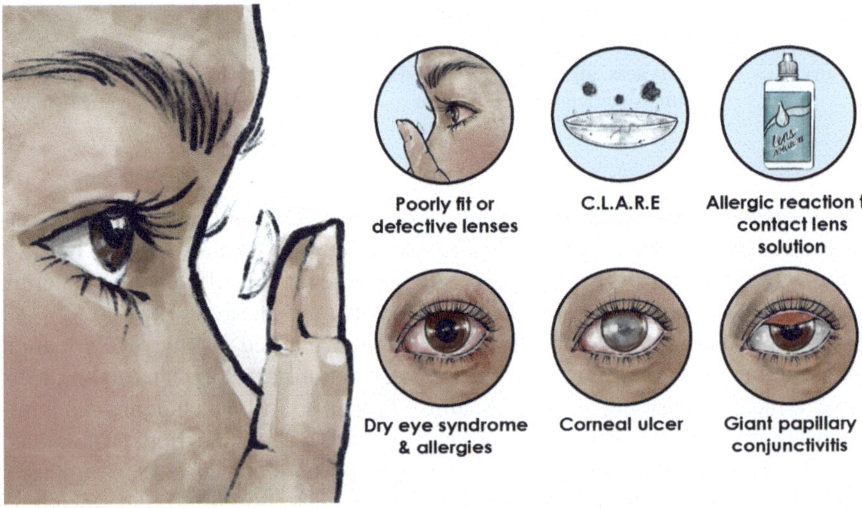

Fig. 6.9 Common reasons contact lenses cause red eyes, including dryness, improper fit, extended wear, hygiene issues, allergic reactions, giant papillary conjunctivitis and CLARE

discomfort require immediate professional attention. Proper lens care, regular follow-ups with an eye care provider, and adherence to replacement schedules are crucial for ensuring safe and comfortable use of contact lenses. Recognizing symptoms early and seeking timely advice can prevent complications, allowing wearers to maintain optimal eye health and vision safety [24].

6.3.3 Understanding Potential Complications: From Discomfort to Serious Eye Conditions

Contact lenses, while highly effective for vision correction, can lead to a range of complications if not used properly. These issues can stem from mechanical trauma, reduced oxygen supply to the cornea due to the lens barrier, or inadequate wetting of the ocular surface. Moreover, poor hygiene, extended wear schedules, and pre-existing conditions can exacerbate these risks, leading to infections, inflammation, or structural damage. Proper care and adherence to recommended practices are critical to mitigating these risks and maintaining eye health. Figures 6.9 through 6.12 illustrate a range of complications, from surface-level damage to more severe conditions, underscoring the importance of early detection and intervention.

One of the most frequently encountered complications is **corneal epithelial defects**, shown in Fig. 6.10a. These defects occur when the outermost layer of the cornea sustains localized damage due to mechanical trauma, such as scratches from fingernails or contact lens edges, or exposure-related factors like dryness and infections. They disrupt the cornea's protective barrier, increasing susceptibility to infections or ulcers [17]. Additional causes, including neurotrophic disease, Bell's palsy, and Sjögren's syndrome, can impair epithelial regeneration and exacerbate the condition. Addressing these underlying factors and adhering to proper lens hygiene and handling practices are essential for effective management [16]. In some cases, long-term complications, such as limbal stem cell deficiency or damage from topical anesthetic abuse, may require more intensive treatments to restore corneal health [18].

Building on this, **corneal edema**, illustrated in Fig. 6.10b, represents another serious complication. This condition involves fluid accumulation within the corneal layers, often resulting from endothelial cell dysfunction caused by trauma, prolonged lens wear, or conditions such as Fuchs' dystrophy. Its hallmark signs include diffuse swelling and a hazy appearance, leading to impaired corneal transparency and blurred vision. Initial treatment typically involves hypertonic saline drops to reduce fluid buildup, but in severe cases, surgical interventions such as Descemet's Stripping Automated Endothelial Keratoplasty (DSAEK) may be necessary to restore normal corneal function [16]. These complications highlight the delicate balance required in contact lens use, reinforcing the importance of regular monitoring and early intervention.

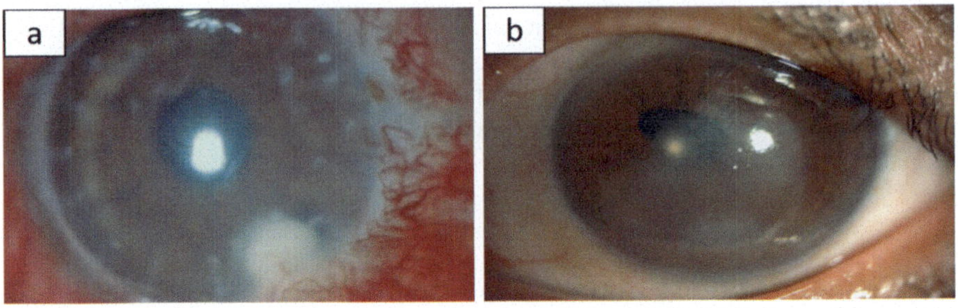

Fig. 6.10 **a** Corneal epithelial defects showing damage to the cornea's outermost layer (reused from Mogado et al. [26], licensed under CC BY4.0); **b** Corneal edema with swelling and hazy transparency due to fluid buildup, impairing iris visibility (reused with permission from Dua and Said [27])

Prolonged use of contact lenses can lead to various structural changes in the cornea, including **corneal distortion**, as shown in Fig. 6.11. This condition, characterized by irregular astigmatism and distorted vision, often results from poorly fitted lenses or delays in replacement schedules. Corneal distortion disrupts the smooth curvature of the cornea, affecting its refractive properties and leading to visual discomfort. Preventing this complication requires ensuring proper lens fit, adhering to recommended replacement intervals, and scheduling regular check-ups to monitor corneal health [28].

In addition to structural changes, inflammatory conditions such as **deep corneal infiltrates** can develop, as illustrated in Fig. 6.12a. These localized inflammatory spots within the cornea, often surrounded by swelling, are commonly associated with contact lens overuse, poor hygiene, or trauma. While non-infectious, sterile infiltrates can mimic infectious ones, making thorough clinical evaluation crucial for accurate diagnosis. Treatment typically involves discontinuing contact lens use and employing anti-inflammatory or antibiotic therapies, depending on the severity and underlying cause [].

A more severe complication, **microbial keratitis**, is depicted in Fig. 6.12b. This aggressive corneal infection, caused by bacteria, fungi, or parasites, presents with symptoms such as intense redness, corneal opacity, and severe pain. Poor contact lens hygiene and exposure to contaminated water are major risk factors for this condition. Acanthamoeba keratitis, a subtype caused by waterborne pathogens, is particularly concerning as it can lead to permanent vision loss if untreated. Early diagnosis and targeted antimicrobial therapy are essential to manage microbial keratitis and prevent irreversible damage [16, 22].

Finally, **corneal neovascularization**, shown in Fig. 6.12c, highlights the long-term consequences of prolonged oxygen deprivation caused by contact lenses with inadequate permeability. This condition results in the growth of abnormal blood vessels into the cornea, compromising its transparency and leading to vision impairment. Management focuses on improving oxygen supply to the cornea, addressing underlying conditions, and discontinuing problematic lenses. In severe cases, surgical

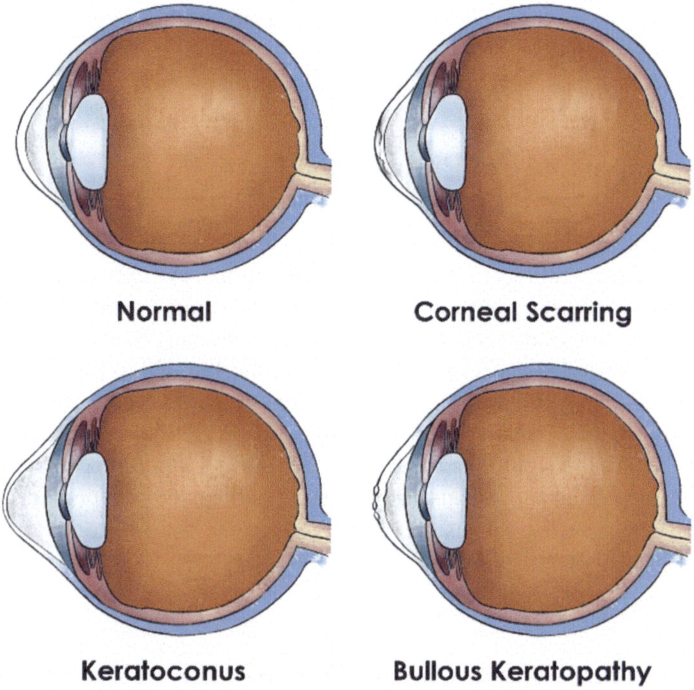

Fig. 6.11 Visualization of cross-sectional of patient with corneal conditions

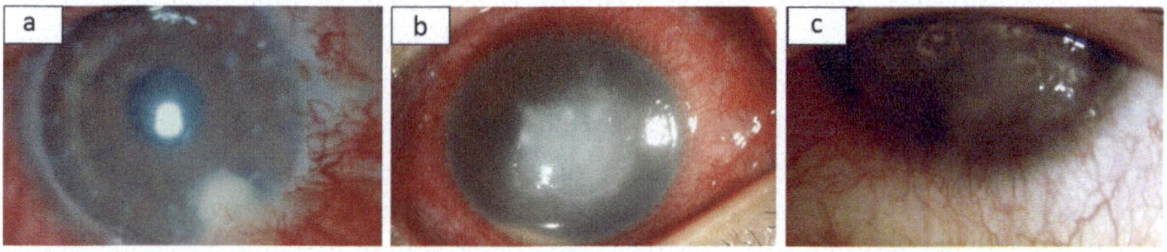

Fig. 6.12 **a** The condition of sterile inflammatory with grey spots surround the cornea (reused from Gracitelli et al. [30] licensed under CC BY4.0). **b** Eye condition of a patient diagnosed with microbial keratitis (reused from Kuo et al. [31], licensed under CC BY4.0). **c** Image of eye with corneal neovascularization (reused with permission from Roshandel et al. [32])

intervention may be required to preserve corneal health and prevent permanent damage [29].

The clinical presentations in Figs. 6.9, 6.10 , 6.11 and 6.12 emphasize the diverse and potentially severe complications associated with improper contact lens use. These cases underscore the critical importance of adhering to proper lens care practices, maintaining regular consultations with eye care professionals, and promptly addressing early symptoms. By following these guidelines, contact lens wearers can minimize risks, manage complications effectively, and ensure long-term ocular health and vision safety.

6.4 The Importance of Professional Care

Professional care is essential for ensuring safe, effective, and comfortable contact lens use. From initial fittings to managing complications, eye care professionals play a critical role in tailoring contact lenses to meet the unique needs of each patient. Regular consultations and comprehensive eye exams not only help detect and address potential issues early but also ensure lenses are properly fitted to prevent discomfort or long-term damage. In addition, managing lens-related complications—whether caused by improper hygiene, wear schedules, or allergies—requires professional guidance to select the right lens modality, care routine, and replacement schedule. By maintaining regular follow-ups and educating patients on best practices, eye care professionals help safeguard vision health, reduce the risk of complications, and promote an overall positive lens-wearing experience.

6.4.1 When and Why to Consult Your Eye Care Professional Guidance

Obtaining a suitable contact lens prescription is essential for maintaining both eye health and clear vision. Contact lenses are medical devices, and like any medical intervention, they require professional oversight. A poorly fitted lens or one made from unsuitable materials can lead to distorted vision, discomfort, infections, inflammation, swelling, or even corneal abrasions. In severe cases, improper lens use may result in permanent damage to the eye. This risk is further exacerbated by sharing lenses, including colored or theatrical ones, which can cause serious, sight-threatening complications. Regular eye exams are vital for ensuring proper lens fit and hygiene, both of which are critical for preventing issues and preserving long-term eye health.

An eye examination involves a series of comprehensive tests conducted by an optometrist, ophthalmologist, or orthoptist to evaluate vision and overall eye health. These tests can detect early signs of serious conditions such as macular degeneration, glaucoma, and diabetic retinopathy. By assessing how well the eyes focus and function together, eye exams provide the foundation for proper vision correction. Essential tests like retinoscopy, refraction, and cover tests ensure early detection of problems, reducing the likelihood of vision loss or complications.

A professional contact lens fitting goes beyond prescription measurements, focusing on safety and comfort. Key evaluations include corneal curvature using a keratometer, corneal topography for more precise measurements, and tear film assessments to address dry eye concerns. Individuals with astigmatism may require toric lenses, which accommodate irregular corneal curvature for a better fit. Following these evaluations, trial lenses are fitted to assess their movement, comfort, and visual performance. Proper fit ensures lenses sit securely without causing irritation or compromising vision.

Follow-up appointments are just as important as the initial fitting. These visits allow professionals to monitor lens performance and address any signs of discomfort, dryness, or irritation. Adjustments can be made to improve fit or comfort, and patients can ask questions or raise concerns about their lenses. Regular follow-ups ensure that contact lens wear remains safe and effective, promoting long-term eye health and reducing risks associated with improper use.

6.4.2 Addressing and Managing Complications with Eye Care Professional

Effective management of contact lens complications begins with choosing the right lens type and care routine. RGP lenses, for example, offer significant advantages for patients who struggle with lens hygiene or wear schedules. These rigid lenses are easier to clean and allow for greater tear exchange, reducing the risk of bacterial buildup and infection [33]. RGP lenses are also well-suited for extended or continuous wear, offering lower complication rates compared to soft lenses. Although some patients initially find them less comfortable, proper education on the adaptation process can ease this transition, allowing them to confidently benefit from the advantages of RGP lenses.

Selecting the right lens modality is crucial for minimizing complications. Daily disposable lenses, which are discarded after each use, are the safest option as they eliminate the risk of debris and bacterial accumulation. Extended wear lenses, though convenient, require careful adherence to cleaning protocols to avoid infections. Patients with dry eyes, allergies, or sensitivities often benefit from daily disposables, as they minimize bioburden and irritation associated with longer wear cycles [12, 34]. For patients with dry eyes, extended wear lenses can exacerbate symptoms due to tear stagnation, leading to discomfort [35]. Daily disposables reduce these risks, ensuring better comfort and reduced complications [36]. Educating patients about these options ensures they choose the modality that best suits their lifestyle and reduces the risk of non-compliance. Additionally, keeping a backup pair of glasses ensures patients have alternatives during allergy season, illness, or when taking breaks from lenses.

While it may be tempting to use store-brand solutions, these often contain older formulations that were designed for pHEMA lens materials, not newer silicone hydrogel lenses. Using these solutions with modern lenses can result in toxic reactions due to preservatives that may irritate the eyes, causing redness and discomfort.

To avoid these issues, it's best to provide patients with a sample of the recommended solution and ask them to return in about a week for a follow-up. This approach helps identify and resolve any issues before they escalate [37].

Regular follow-up visits are essential for maintaining healthy contact lens wear. Allergic reactions to preservatives in lens solutions or lenses themselves can occur quickly, especially if the patient has been sensitized to certain ingredients. New users may take months to experience adverse reactions, so multiple follow-ups are necessary during the first few months of wear. If no issues arise during this period, the patient can transition to annual visits. It's crucial to stress the importance of annual eye exams, as both ocular and systemic changes can affect contact lens wear over time, and regular check-ups are necessary to ensure continued comfort and safety.

Deciding between daily disposable lenses and extended wear lenses is essential for patient comfort and safety. By the second follow-up visit, which typically occurs within the first month of wear, practitioners can assess whether a patient is suited for extended wear or short-cycle lenses. For younger patients, especially teenagers, it's advisable to start with shorter wear cycles, as they tend to be less compliant with wear schedules. Ensuring patients follow proper wear schedules significantly reduces the risk of complications. By matching patients with the appropriate lens modality, practitioners can mitigate non-compliance and avoid issues stemming from improper lens care.

Educating patients on proper contact lens wear and care is crucial for preventing complications and ensuring long-term eye health. Patients should always follow their doctor's instructions on wearing schedules and proper handling techniques. Ongoing education helps emphasize the importance of compliance and hygiene, reducing risks such as infections or dry eyes. A structured patient education plan should include the following key points:

(a) Wearing Schedule: Clearly outline the recommended wearing schedule per day and per week, tailored to the patient's needs and lens type.
(b) Recognizing and Managing Problems: Educate patients about potential issues they might encounter, such as redness, discomfort, or blurred vision. Provide clear guidance on what to do in these situations, including when to seek professional help.
(c) Replacement Schedule: Stress the importance of adhering to the recommended replacement schedule for both lenses and lens cases to prevent contamination and wear-related issues.
(d) Proper Disinfection Practices: Explain the recommended disinfection system in detail, including step-by-step instructions on its correct usage. Highlight the critical role of lens disinfection in maintaining eye health and tailor the disinfection method to the patient's specific needs.
(e) Follow-Up Appointments: Set clear expectations about when and why the patient should return for follow-ups. Reinforce the importance of these appointments in monitoring eye health and updating care routines.

This comprehensive approach empowers patients to manage their contact lenses responsibly, ensuring long-term comfort, safety, and vision clarity. Regular communication between patients and eye care professionals remains central to addressing complications and optimizing lens wear [13].

References

1. Lakkis, C., Lorenz, K.O., Mayers, M.: Topical review: contact lens eye health and safety considerations in government policy development. Optom. Vis. Sci. **99**, 737–742 (2022). https://doi.org/10.1097/OPX.0000000000001937
2. Vianya-Estopa, M., Ghorbani-Mojarrad, N., Huntjens, B., Garcia-Porta, N., Piñero, D.P., Nagra, M., Terry, L., Dutta, D., Wolffsohn, J., Joshi, M., Martin, E., Maldonado-Codina, C.: Current approaches to soft contact lens handling training—global perspectives. Contact Lens Anterior Eye **46**, (2023). https://doi.org/10.1016/j.clae.2023.102068
3. American Optometric Association: Comprehensive adult eye and vision examination. Clin. Pract. Guidel. **51**, 24–30 (2015)
4. David Tubert: Eye Exam and Vision Testing Basics When Should You Have an Eye Exam ? What Do Ophthalmologists Check. 1–4 (2024). https://www.aao.org/eye-health/tips-prevention/eye-exams-101
5. Schornack, M.: Prescription and management of contact lenses in patients with monocular visual impairment. Optometry **78**, 652–656 (2007). https://doi.org/10.1016/j.optm.2007.02.022
6. Clarity Vision: Do I Need Another Eye Exam For Contacts Than For Glasses?, https://www.findclarityvision.com/blog/is-an-eye-exam-for-contacts-and-glasses-the-same.html
7. EyeRis, Vision, E.: Soft Contact Lens Guide, https://www.eyerisvision.com/soft-contact-lens-guide.html
8. McMonnies, C.W.: Hand hygiene prior to contact lens handling is problematical. Contact Lens Anterior Eye **35**, 65–70 (2012). https://doi.org/10.1016/j.clae.2011.11.003
9. Alensa: How to put in and put out contact lenses—Guide|Alensa, https://www.alensa.com/information/contact-lens-application.html
10. Michigan et al: Essential Effective Guide To Cleaning Contact Lenses, https://michigancontactlens.com/how-to-clean-contact-lenses/
11. CDC: About Cleaning, Disinfecting, and Storing Contact Lenses|Healthy Contact Lens Wear and Care|CDC, https://www.cdc.gov/contact-lenses/about/about-cleaning-disinfecting-and-storing-contact-lenses.html
12. Rhee, M.K., Jacobs, D.S., Dhaliwal, D.K., Szczotka-Flynn, L., Prescott, C.R., Jhanji, V., Steinemann, T.L., Koffler, B.H., Jeng, B.H.: Contact lens safety for the correction of refractive error in healthy eyes. Eye Contact Lens **48**, 449–454 (2022). https://doi.org/10.1097/ICL.0000000000000938
13. Lievens, C.W., Cilimberg, K.C., Moore, A.: Contact lens care tips for patients: an optometrist's perspective. Clin. Optom. **9**, 113–121 (2017). https://doi.org/10.2147/OPTO.S139651
14. Cooper Vision: Removing Your Contact Lenses | CooperVision, https://coopervision.com/new-wearer/removing-your-contact-lenses
15. Nichols, J.J.: Evaporative tear film and contact lens factors associated. Ohio State Univ. (2004)
16. Alipour, F., Khaheshi, S., Soleimanzadeh, M., Heidarzadeh, S., Heydarzadeh, S.: Contact lens-related complications: a review. J. Ophthalmic Vis. Res. **12**, 193–204 (2017). https://doi.org/10.4103/jovr.jovr_159_16
17. Gurnani, B., Kaur, K.: Contact lenses. Br. Med. J. **288**, 1–16 (2023). https://doi.org/10.1136/bmj.288.6417.586
18. Juhong, J., Mordmuang, A., Jewboonchu, J., Rattanathamma, P., Narkkul, U., Karnjana K., Udomwech, L. Rub and Rinse Contact lenses before wearing as a protective regimen against

contact lens-related eye infections. Clin Ophthalmol. **16**, 567–577 (2022). https://doi.org/10.2147/OPTH.S357099
19. Kates, M.M., Tuli, S.: Complications of contact lenses. JAMA J. Am. Med. Assoc. **325**, 1912 (2021). https://doi.org/10.1001/jama.2020.20328
20. Lim, C.H.L., Stapleton, F., Mehta, J.S.: Review of contact lens–related complications. Eye Contact Lens **44**, S1–S10 (2018). https://doi.org/10.1097/ICL.0000000000000481
21. Stapleton, F., Bakkar, M., Carnt, N., Chalmers, R., Vijay, A.K., Marasini, S., Ng, A., Tan, J., Wagner, H., Woods, C., Wolffsohn, J.S.: CLEAR—Contact lens complications. Contact Lens Anterior Eye **44**, 330–367 (2021). https://doi.org/10.1016/j.clae.2021.02.010
22. Waghmare, S.V., Jeria, S.: A review of contact lens-related risk factors and complications. Cureus. **14**, 1–8 (2022). https://doi.org/10.7759/cureus.30118
23. Wong, J.C., Barak, A.: Managing dry eye disease with novel medications: mechanism, study validity, safety, efficacy, and practical application. Pharmacy. **12**, 19 (2024). https://doi.org/10.3390/pharmacy12010019
24. Your Eye Site, S., Appointment, R.A., Solutions, T.: 7 Common Contact Lens Problems and Their Solutions, pp. 2–5 (2021)
25. Very Well Heatlh: Causes of Red Eyes in Contact Wearers, https://www.verywellhealth.com/red-eyes-and-contacts-3421657
26. Morgado, R.V., Moura, R., Moreira, R., Falcão-Reis, F., Pinheiro-Costa, J., Morgado, R.V., Moura, R., Moreira, R., Falcão-Reis, F., Sr., J.P.-C.: New promising therapeutic approach for refractory corneal epithelial defects. Cureus. **15**, (2023). https://doi.org/10.7759/CUREUS.39324
27. Dua, H.S., Said, D.G.: Clinical evidence of the pre-Descemets layer (Dua's layer) in corneal pathology. Eye **30**(8), 1144–1145 (2016). https://doi.org/10.1038/eye.2016.62
28. East Valley Ophthalmology: Cornea Specialists | East Valley Ophthalmology Eye Doctors Mesa AZ, https://doctor-hill.com/mesa-eye-doctors/specialists/cornea/
29. Ozdemir et al.: Analysis of the horizontal corneal diameter, central corneal thickness, and axial length in premature infants, https://www.scielo.br/j/abo/a/LW4Rg4fzgrrqgLKVbVsjWTb/
30. Gracitelli, B., Ferrar, P.V., Alberto, C., Pereira, P., Freitas, D.D.: A case of recurrent keratitis caused by Paecilomyces lilacinus and treated by voriconazole Uso do voriconazol na ceratite recorrente causada por fungo Paecilomyces lilacinus. **82**, 152–154 (2019)
31. Kuo, M.T., Hsu, B.W.Y., Lin, Y.S., Fang, P.C., Yu, H.J., Chen, A., Yu, M.S., Tseng, V.S.: Comparisons of deep learning algorithms for diagnosing bacterial keratitis via external eye photographs. Sci. Rep. **11**, 1–10 (2021). https://doi.org/10.1038/s41598-021-03572-6
32. Roshandel, D., Eslani, M., Baradaran-Rafii, A., Cheung, A.Y., Kurji, K., Jabbehdari, S., Maiz, A., Jalali, S., Djalilian, A.R., Holland, E.J.: Current and emerging therapies for corneal neovascularization. Ocul. Surf. **16**, 398–414 (2018). https://doi.org/10.1016/j.jtos.2018.06.004
33. Holden, B.A., Sweeney, D.F., Vannas, A., Nilsson, K.T., Efron, N.: Effects of long-term extended contact lens wear on the human cornea. Invest. Ophthalmol. Vis. Sci. **26**, 1489–1501 (1985)
34. Hayes, V.Y., Schnider, C.M., Veys, J.: An evaluation of 1-day disposable contact lens wear in a population of allergy sufferers. Contact Lens Anterior Eye **26**, 85–93 (2003). https://doi.org/10.1016/S1367-0484(03)00019-5
35. Markoulli, M., Kolanu, S.: Contact lens wear and dry eyes: challenges and solutions. Clin. Optom. **9**, 41–48 (2017). https://doi.org/10.2147/OPTO.S111130
36. Muntz, A., Subbaraman, L.N., Sorbara, L., Jones, L.: Tear exchange and contact lenses: a review. J. Optom. **8**, 2–11 (2015). https://doi.org/10.1016/j.optom.2014.12.001
37. Willcox, M.D.P.: Solutions for care of silicone hydrogel lenses. Eye Contact Lens **39**, 24–28 (2013). https://doi.org/10.1097/ICL.0b013e318275e0d9

Chapter 7
Emerging Technology in Contact Lenses

Abstract This chapter explores the cutting-edge advancements shaping the future of contact lenses, showcasing their transition from vision correction devices to multi-functional tools for healthcare and beyond. It begins by highlighting the potential of next-generation lenses, such as smart contact lenses that integrate sensors, micro-electronics, and augmented reality (AR) displays, enabling real-time health monitoring and enhanced visual experiences. The chapter delves into the advancements in nanotechnology and innovative materials that are revolutionizing lens design, improving comfort, wearability, and manufacturing precision. Healthcare applications take center stage as the chapter examines lenses designed for myopia control, drug delivery, and continuous health monitoring, demonstrating how contact lenses are becoming vital tools in personalized medicine. Additionally, the chapter addresses the challenges and opportunities in developing these emerging technologies, such as regulatory hurdles, cost barriers, and user adaptation. In reflecting on these breakthroughs, the chapter underscores the challenges and opportunities ahead—balancing technological complexity with user accessibility and navigating regulatory landscapes while maintaining affordability. As the journey culminates, the transformative potential of contact lenses stands as a testament to the limitless possibilities of innovation, offering a glimpse into a future where the boundaries of vision and technology are reimagined.

Keywords Smart contact lenses · Nanotechnology · Augmented reality lenses · Healthcare applications · Drug delivery · Innovative materials · Wearable technology · Vision enhancement · Future trends

7.1 Beyond Vision Correction: The Next-Generation Contact Lenses

The contact lens industry has undergone a tremendous transformation in recent decades, driven by advances in material science, manufacturing processes, and a deeper understanding of the human eye. While the primary function of contact

© The Author(s), under exclusive license to Springer Nature Singapore Pte Ltd. 2025

N. Adrus et al., *Contact Lenses*, Series in BioEngineering,
https://doi.org/10.1007/978-981-96-3636-5_7

lenses has traditionally been to correct refractive errors, there is currently exploration into other potential uses of these versatile devices. A wide range of therapeutic and diagnostic capabilities is being considered. Imagine a future where your contact lenses could do more than just improve your vision—they could also monitor your glucose levels, deliver medication for glaucoma, or even provide augmented reality (AR) experiences. One of the most exciting emerging trends in contact lens technology is the integration of drugs and other bioactive compounds. Beyond merely addressing refractive errors, these cutting-edge lenses promise a new era of functionality, blending advanced materials and smart technology to enhance and expand their role in our daily lives. From health monitoring to AR, the future of contact lenses is not just about clear vision—it's about a world of possibilities.

The evolution of smart contact lenses has been marked by several transformative milestones, as outlined in Fig. 7.1. The foundation for these advancements was established between 2003 and 2009, beginning with the development of intraocular pressure (IOP) monitoring lenses [1]. This early progress led to the 2009 launch of the **Triggerfish® contact lens** by SENSIMED AG, a silicone-based lens featuring an embedded antenna for continuous IOP monitoring [2]. By 2010, the focus shifted toward wireless power and display technologies, with significant contributions by Lingley et al. [3]. The year 2011 also saw the introduction of glucose-sensing lenses by Yao et al. [4], designed for continuous, non-invasive health monitoring.

Between 2014 and 2016, Google partnered with Alcon, Novartis's eye-care division, to explore glucose-monitoring lenses, though this project faced challenges due to the limited correlation between tear and blood glucose levels. In parallel, the Triggerfish® lens received Food and Drug Administration (FDA) clearance in 2016, solidifying its utility for 24-h IOP monitoring [5, 6]. Material innovations in 2016, such as the integration of graphene electrodes by Kaur et al. [7], demonstrated progress toward adaptable focus mechanisms for presbyopia. Recent advancements from 2019 onward have ushered in the bioelectronic era, with technologies such as supercapacitor-integrated lenses [8] and artificial iris systems [9]. In 2024, self-lubricating lenses, developed to address dry eye syndrome, further enhanced user comfort and functionality. Collectively, these milestones illustrate the continuous innovation driving the development of smart contact lens, spanning health monitoring, material science, and wearable bioelectronics.

Contact lenses are now a well-established yet constantly evolving technology, driven by the growing demand for it continues to increase. The global contact lens market size was valued at USD 10.45 billion in 2023. The market is projected to grow from USD 11.08 billion in 2024 to USD 18.30 billion by 2032, exhibiting a compound annual growth rate (CAGR) of 6.5% during the forecast period [10]. With continued development of its technology, the materials used in contact lenses have also evolved, with polymers tailored to address the limitations about ocular complications induced by contact lens wear. The choice of the polymer whether for rigid gas-permeable (RGP) in ocular therapeutics or soft contact lenses for vision correction is carefully guided by application needs.

The evolution of smart contact lenses reflects a remarkable journey of multidisciplinary collaboration across material science, microelectronics, and healthcare.

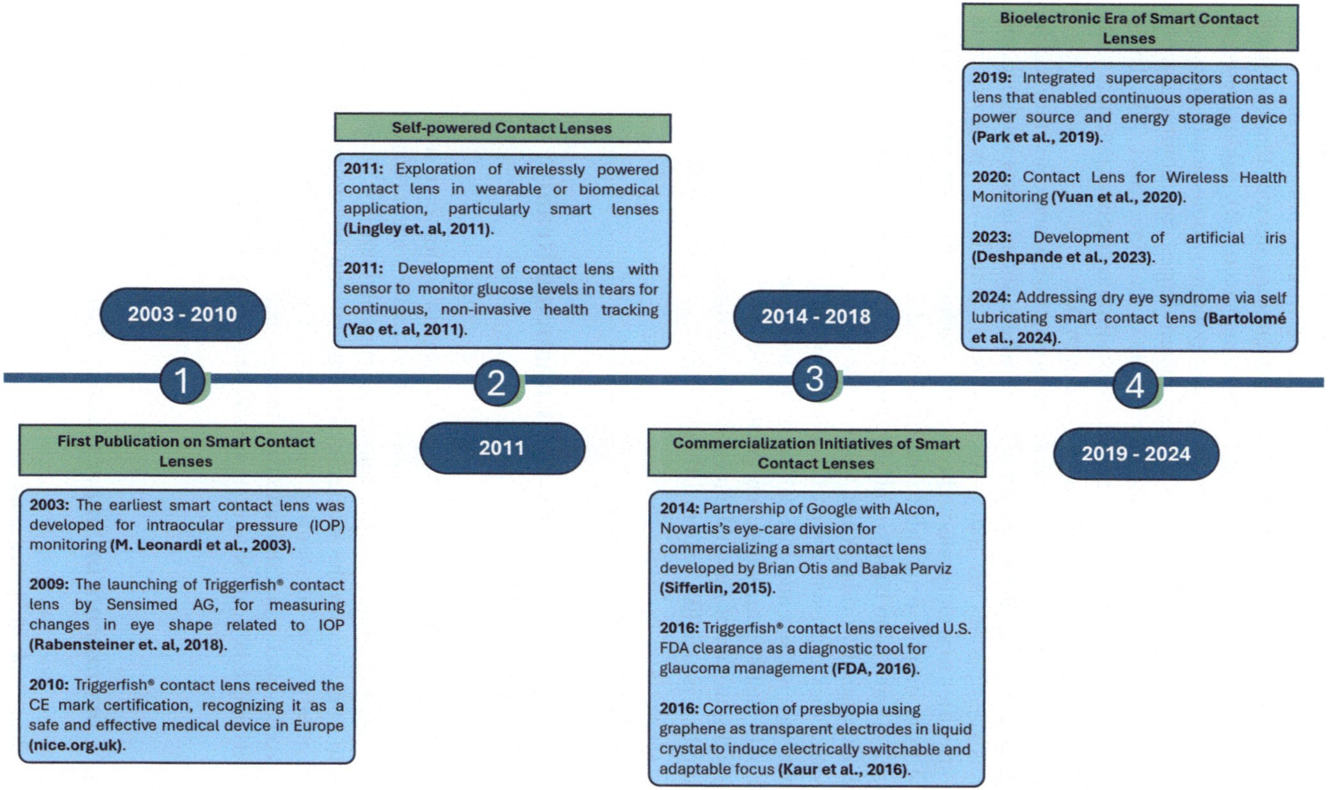

Fig. 7.1 Timeline of smart contact lens development highlighting key milestones: (1) 2003–2010: Early IOP-monitoring lenses and the launching and certification of Triggerfish® contact lens; (2) 2011: Wireless power and glucose-sensing lenses; (3) 2014–2018: Commercialization initiatives and advancements in material innovation; (4) 2019–2024: Bioelectronic era featuring supercapacitors, artificial iris systems, and self-lubricating lenses

From the initial vision of integrating electronics into eyewear to the latest prototypes capable of providing real-time health monitoring and AR, the potential of smart lenses is vast and exciting. Miniaturized sensors, circuits, and wireless power systems now allow these lenses to go beyond vision correction, creating devices that can track eye movement and adapt to changing light conditions (Sect. 7.2). Additionally, as these intelligent lenses are designed to detect and respond to various health markers, they offer potential for personalized healthcare, where lenses can monitor glucose level and even respond to changes in the user's body in real time (Sect. 7.3). These capabilities mark a new frontier for contact lenses, positioning them as potential tools for health monitoring, therapeutic treatment, and even AR.

While challenges remain in technology, regulatory approval, and consumer acceptance, the groundwork has been laid for a future where smart lenses may become as common as smartphones are today. As these intelligent devices progress from laboratory curiosities to widely adopted consumer products, they are set to redefine the transition from simple visual aids to multifunctional interfaces. These interfaces not only augment our vision but also enhance our lives, offering a seamless blend of the physical and digital worlds. While we wait for smart contact lenses to become a common sight, they represent a significant leap forward in wearable tech, poised to change the way we see and interact with the world around us [11, 12]. Their journey has only just begun, with even greater possibilities on the horizon. For a more in-depth discussion of the challenges and future directions in smart lens technology—including issues like battery life, data privacy, and the path to consumer acceptance—please refer to Sect. 7.4.

7.2 Integrating Functionality and Enhancing Reality with Nanotechnology

Smart contact lenses represent a revolutionary leap in vision care and wearable technology. Unlike traditional contact lenses that primarily correct refractive errors, smart lenses incorporate miniaturized electronic circuits and sensors. This integration allows them to interact with the wearer's environment and physiology, gathering data and potentially even augmenting their vision in real-time. Recent advances in microelectronics and nanofabrication on flexible substrates have now enabled sensors, circuits and other essential components to be integrated on a small contact lens.

7.2.1 The Components in Smart Contact Lenses

According to Xia et al. [13], the smart contact lens is typically composed of six building blocks: sensing, energy management, driver electronics, communications, substrate and the interfacing modules as depicted in Fig. 7.2. Each module plays an essential role in enabling the complex functions of the lens, from health monitoring to AR overlays.

As shown in Fig. 7.2, the **Energy module** is responsible for power harvesting and storage, ensuring a stable energy source to support continuous operation. The **Driver Chip module** manages and regulates this energy, providing a reliable power flow to all electronic components and ensuring safe operation across modules. Furthermore, the **Substrate & Interconnection module** hosts and links these electronic modules, creating a cohesive structure that facilitates communication and power distribution among all embedded components.

The **sensing module** is critical for collecting physiological and environmental data, enabling real-time insights. Capacitive sensors, for example, are mostly used to measure the IOP of the eye in smart contact lenses. This concept involves placing a dielectric layer between two electrodes that conforms to the curvature around the cornea. The compression of the dielectric layer due to increased IOP leads to a rise in capacitance, which can then be measured to monitor pressure changes. Notable examples include the work by Zhu et al. [14] and Yang et al. [15], who developed smart contact lenses with a wireless capacitive sensing design for IOP monitoring. Specifically, Zhu et al. [14] addressed the hydrogel swelling in hydrogel-based smart lenses by using a conformal stacking technique with an embedded pyramid-microstructure capacitive sensor. Another approach was reported by Bobrowski et al. [16], who

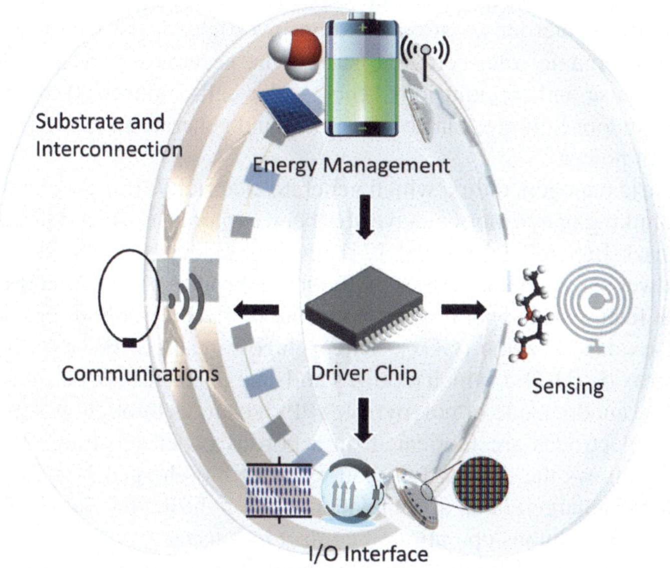

Fig. 7.2 Building blocks of smart contact lenses, which include the energy management, communications, driver chip, sensing, input/output (I/O) interface and the substrate & interconnection modules (reproduced with permission from Xia et al. [13], licensed under CC BY 4.0)

proposed hybrid circuitry that functions as both biofuel cells and capacitors, optimizing the limited space within lens. This self-charging biosupercapacitor harnesses energy from tear glucose by incorporating indium tin oxide nanoparticles to enhance capacitive and catalytic currents on electrode surfaces. Additionally, strain sensors measure IOP and work by detecting mechanical distortion within the lens material, converting these disruptions into electrical signals for monitoring [17]. These sensors are usually made from conductive and flexible materials, such as metallic nanowires [18, 19] and graphene [20, 21].

The **Energy Management** or **Power Source** plays a critical role in the demands of continuous monitoring and data transmission, with smart contact lenses fitted with complex circuity to fulfill these energy requirements. Solid-state batteries have been designed to be thinner and more flexible, allowing for direct implementation on curved polymer surfaces. These batteries provide a constant power supply, enabling predictable and consistent usage, whereas wireless power transmission may experience disconnection if the device moves away from the path of the energy source.

Innovative approaches to energy management include supercapacitors and biofuel cells. For example, Park et al. [8] used silicon diodes to charge a supercapacitor within a smart contact lens, providing a stable DC input essential for reliable performance. Lee et al. [22] further advanced this approach by manufacturing solid-state lithium-ion batteries, enhancing power density within the compact design of smart lenses.

Biofuel cells offer a unique, self-sustaining power source by converting energy from biological substances into electricity. These cells can be fabricated directly onto hydrogel surfaces through a single step annealing process [22]. Novel methods of powering smart contact lenses are also being investigated. For example, Lin et al. [23] developed organic solar cells that supply the necessary power for biosensors measuring glucose and calcium ions. These solar cells, optimized to operate efficiently under indoor lighting, can absorb protos and generate an electrical current of about 0.6 V of power.

Triboelectric nanogenerators, which generate electricity from mechanical movements like blinking or eye motion, have also been explored as a novel power solution for smart lenses. In parallel, Yun et al. [24] developed an enzyme-based bio-charging design to power smart contact lenses, without needing an external electrical supply. Their design features a copper hexacyanoferrate (CuHCFe) cathode combined with the glucose oxidase enzyme (GOx). When glucose reacts with GOx, it generates hydrogen peroxide (H_2O_2), which oxidizes and charges the CuHCFe cathode. At the same time, the anode, made of polypyrrole (PPy), charges through its self-reduction ability. Both electrodes are fabricated on a porous paper substrate. This innovative approach allows the contact lens battery to be bio-charged in a glucose-based overnight and discharged in an artificial tear solution during the day, simulating the daily cycle and continuous operation of contact lens usage.

These advancements highlight the ongoing development in power management for smart contact lenses, focusing on optimizing both energy harvesting and power storage. These efforts aim to achieve self-sustainable and autonomous contact lenses, for smarter and more practical wearable devices.

The **Driver Chip** is a central part of a smart contact lens, effectively managing and regulating received energy, to ensure all electronic modules operate safely. This module often includes rectifiers, regulators and converters, which control and stabilize power signals throughout the system. Moreover, it processes signals through oscillators, sensor readout circuitry and logic circuitry. This is often required for reading data from the sensing module or sending control signals to the I/O Interface.

The **Communication** module in smart contact lenses is concerned with transmitting and receiving data. For transmission, sensor data is processed by a coder and mixer, with the resulting output signal transmitted via antenna. Smart contact lenses, especially those designed as human–machine interface (HMI) applications, rely heavily on antennas to enable data exchange between the contact lens platform and an external processor. Consequently, most of the development within the communications module is centered on antenna design. Key parameters for antenna design in smart contact lenses include return loss, bandwidth, operating frequency, flexibility and vision obstruction.

To optimize antenna performance with minimum vision obstruction, researchers often use a ring-type antenna geometry of various designs, as shown in Fig. 7.3. These designs maximize antenna size while ensuring seamless integration into the contact lens platform. Figure 7.3a illustrates a segmented ring antenna design with distinct inner and outer loops separated by a gap, which helps create a uniform magnetic field distribution for efficient power transfer [25]. Spiral-style antennas designed by Kim et al. [26] have also emerged as promising geometries, enabling effective data transmission without compromising user comfort or visual clarity (Fig. 7.3b).

Given the essential role of communication in smart contact lenses, these lenses must transmit and receive information seamlessly to interact effectively with external devices, such as smartphones and other medical monitoring devices. This capability enables real-time health monitoring and rapid therapeutic responses, which are particularly crucial for managing chronic conditions like glaucoma or diabetes. Antennas are the mainstay for outgoing communication, and their design are critical in maximizing functionality while ensuring the communication module fulfils its vital role in smart contact lens technology.

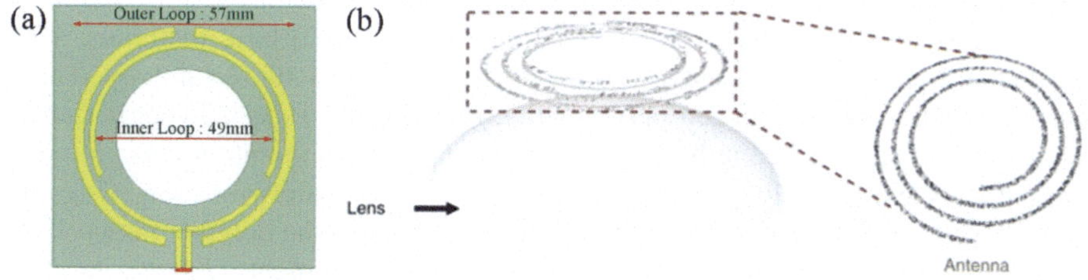

Fig. 7.3 **a** Ring-type antenna designs for smart contact lenses, featuring inner and outer loops (adapted with permission from Chiou et al. [25], licensed under CC BY 4.0), and **b** spiral-style antennas (adapted with from Kim et al. [26], licensed under CC BY 4.0). Both designs emphasize enhanced signal performance and seamless integration

Substrate and Interconnect Technology: All electronic components are fabricated on a particular substrate and interconnected with wires. Flexible and stretchable substrates are essential to ensure user comfort while maintaining compatibility with embedded electronic components. Traditionally, contact lenses are made from polymer and silicone hydrogel-based **substrate** [27], but future designs require enhanced materials that ensure electronic safety and durability. Encapsulation materials are also under investigation to prevent wearer harm from potential electronic hazards. **Interconnects** in smart contact lens must be transparent, conductive, flexible, stretchable with low sheet resistance to maintain performance and comfort. In this context, the futuristic concept of smart contact lenses, integrating all main components and their interactions, should be viewed as in Fig. 7.2. This figure would depict how biosensors, drug delivery systems, and monitoring capabilities work together to enhance the diagnosis and treatment of various eye disease, highlighting the potential for real-time data collection and improved patient outcomes.

The **I/O interface** serves as the critical bridge between the advanced technologies embedded in smart contact lenses and the external world. By leveraging state-of-the-art substrates and interconnects, the I/O interface ensures seamless interaction, enabling lenses to relay data from biosensors while also providing real-time feedback through AR overlays. This capability transforms AR lenses into powerful tools for immersive visual experiences, navigation, and health monitoring. As such, the I/O interface stands as a cornerstone of AR applications in smart contact lenses, connecting innovative materials and electronics with user-centric functionality.

7.2.2 Nanotechnology in Contact Lens Design and Materials

Recent advancements in **nanotechnology** and **material science** have revolutionized the design and functionality of contact lenses, leading to the development of nanocomposite **materials** that enhance comfort, wearability, and therapeutic capabilities. These innovations build on advancements in base polymers, functional coatings, and sustainable materials, ensuring improved performance across diverse applications.

Base Polymer Advances: The development of advanced base materials, such as silicone hydrogels, represents a major leap in contact lens technology. These materials offer superior oxygen permeability, reducing the risk of hypoxia during extended wear [28]. Silicone hydrogels are also tailored to maintain lens softness, transparency, and durability, ensuring optimal comfort and safety for users. Ongoing advancements in polymer chemistry have further optimized the polymer networks to allow for longer wear durations, even in challenging environments (*Further reading in Sect.* 2.3).

Additionally, the **hybridization of polymers with nanoparticles** has further enhanced contact lens properties [29]. While soft hydrogel contact lenses with high water content typically exhibit a lower refractive index (RI) compared to RGP lenses, hybrid nanocomposites bridge this gap. Although RGP lenses correct vision effectively and feature higher RI, their low water content limits comfort. Recent innovations have leveraged titanium dioxide (TiO_2) nanoparticles in poly(methyl methacrylate) PMMA-based RGP lenses to enhance optical performance and structural integrity [29]. These nanocomposites produce highly transparent lenses (>90% light transmittance), maintaining the rigidity needed for effective vision correction while improving flexibility and reducing optical aberrations. Additionally, neodymium oxide—a safer alternative to silver nanoparticles—has been incorporated into hydrophilic lens materials to enhance tensile strength and wettability [30]. These advancements not only optimize lens performance but also ensure user comfort and extended wearability in both soft and rigid lens designs.

The integration of **enhanced vision properties** has been another focus of research. Nanoparticles embedded into lenses can modify their optical properties, leading to improved vision correction and enhanced visual acuity [31]. For instance, the inclusion of TiO_2 nanoparticles with specific refractive indices within the lens material such as PMMA in PMMA contact lenses progressively enhances the clarity and sharpness of retinal images. While traditional PMMA lenses offer basic correction (Fig. 7.4a, b, the nanocomposite lenses (Fig. 7.4c, d) provide a notable improvement, with higher nanoparticle concentrations yielding clearer image resolutions [29]. Furthermore, **photochromic nanocomposite lenses** exhibit tint changes in response to UV or visible light, protecting users against harmful radiation and glare [32, 33]. These features are particularly beneficial for outdoor activities and environments with varying lighting conditions, significantly improving user comfort and visual performance [34].

Functional Polymers and Coatings: Incorporating functional polymers into lens materials has enabled novel properties, such as stable hydration and resistance to protein adhesion. Zwitterionic polymers, for example, electrostatically bind to aqueous components in the eye region, promoting long-lasting hydration and minimizing dryness [35]. Advances in hydrophilic polymers have further improved the wettability of lenses, addressing common discomfort associated with traditional materials [27]. These functional coatings also enhance biocompatibility, making lenses suitable for a wider range of users, including those with sensitive eyes.

Building on these advances, the application of **3D printing with nanomaterials** has redefined customization and precision in lens manufacturing. This technology enables the creation of lenses tailored to individual wearers through digital imaging and scanning techniques. Complex geometries, such as multifocal and toric lenses, can now be produced with embedded functionalities, including sensors and enhanced optical performance [36, 37]. Moreover, **rapid prototyping and efficient material selection** facilitated by 3D printing are accelerating the development of innovative designs while maintaining manufacturing precision and consistency [38]. This ensures optimal biocompatibility, durability, and user comfort.

Environmental Sustainability: In response to growing concerns about the environmental impact of disposable lenses, research into biodegradable and eco-friendly materials has gained momentum. At UTM, our research group is pioneering the development of ***OrthaG***, a bio-based hydrogel derived from agricultural sources such as Orthosiphon aristatus stem fiber and pineapple leaf fiber (Fig. 7.5). Our findings

Fig. 7.4 Retinal image simulation of the human eye: **a** without a contact lens, **b** with pure PMMA contact lens, **c** with 0.005 PMMA-TiO_2 nanocomposite lens, and **d** with 0.01 PMMA-TiO_2 nanocomposite lens, illustrating progressive improvement in image clarity and resolution with increasing TiO_2 nanoparticle concentrations (reused with permission from Shaker et al. [29], licensed under CC BY 4.0)

highlight OrthaG's ability to optimize transparency, water content, and elasticity for contact lenses, making it both eco-friendly and cost-effective. Beyond enhancing user comfort and biocompatibility, this innovation supports sustainable manufacturing practices, benefiting collaborators like Alcon Ciba Vision, Johor, through streamlined production and reduced costs. OrthaG also aligns with Sustainable Development Goal 12, promoting sustainable consumption and production. This research exemplifies a crucial step toward environmentally responsible and high-performing contact lenses, contributing to a more sustainable future in eye care.

As research progresses, the convergence of nanotechnology and advanced materials continues to redefine the potential of contact lenses. These innovations are not only improving vision correction but also paving the way for transformative functionalities that elevate the role of contact lenses in healthcare and interactive technologies. Transitioning seamlessly from enhancing optical properties to enabling advanced capabilities like AR, nanotechnology serves as a cornerstone in the evolution of smart contact lenses.

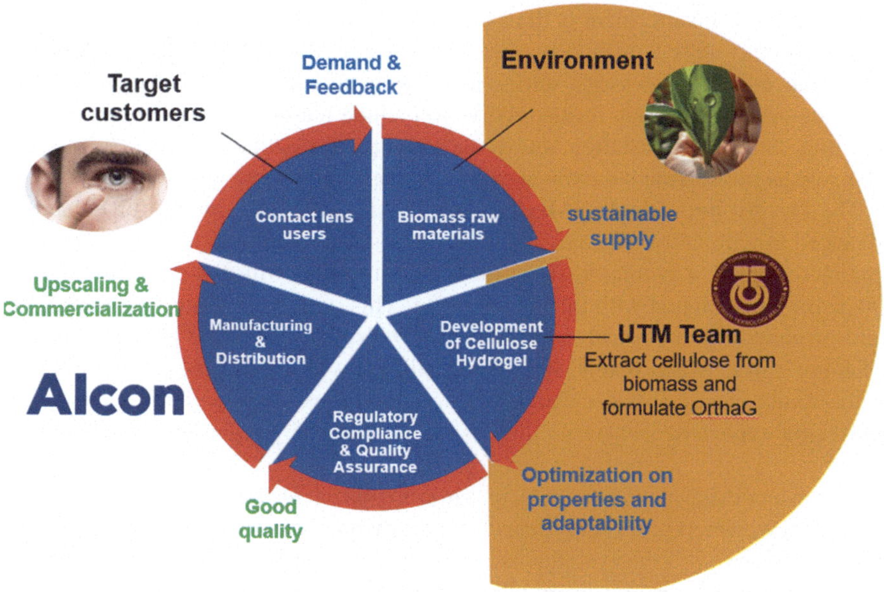

Fig. 7.5 A framework highlighting the development of OrthaG cellulose hydrogel from biomass, led by the UTM team in collaboration with Alcon, focusing on sustainability and optimized material properties for advanced contact lens production

7.2.3 Enhancing Functionality: The Interface and Integrated Augmentation

Building on the foundations of **nanotechnology** and **miniaturized components** discussed in the previous sections, **AR applications** represent a pivotal evolution in the development of smart contact lenses. By leveraging advancements in nanotechnology, such as improved power systems and ultra-thin materials, AR lenses are able to seamlessly integrate advanced functionalities into a compact form. This section explores how these embedded technologies interact to deliver real-time data, AR capabilities, and personalized visual enhancements. By combining ergonomic designs with cutting-edge components, AR lenses transcend traditional vision correction, transforming into dynamic tools for navigation, immersive augmented experiences, and tailored vision augmentation.

AR contact lenses eliminate the need for bulky glasses or headsets, offering users a lightweight and seamless experience. By combining sensors, power management systems, communication modules, and display mechanisms, these lenses redefine the boundaries of traditional eyewear. AR lenses can overlay digital information onto the real world, enabling real-time updates such as navigation, weather information, and notifications. These capabilities allow users to access information effortlessly, enhancing both convenience and functionality.

One notable advancement is the Mojo Vision AR contact lens, released in 2022 [39]. This innovative lens incorporates cutting-edge MicroLED display technology and advanced motion sensors, setting a new benchmark for AR-enabled lenses. In the future, AR lenses may further evolve to include features like optical zoom, text scrolling, language translation, and visual prompts for public speaking, solidifying their role as transformative tools in everyday life.

The **Design and Display Mechanism** of AR lenses are critical for balancing advanced functionality with user comfort. For instance, Mojo Vision's smart contact lens incorporates a groundbreaking display that moves with the eye, ensuring the image remains aligned with the user's vision. This design integrates a compact 15° field-of-view, with the world tiniest MicroLED display, measuring just 480 μm with a 1.8 μm gap between pixels. By staying aligned with the fovea—the central part of the retina responsible for detailed vision—the lens offers a 'built-in' **foveated rendering experience**, ensuring that vision remains sharp regardless of where user looks.

However, design challenges remain. Mojo Vision's lens, for instance, has a limited field of view, which poses a bottleneck for many use cases. While the ability of the display to move with the eye alleviates this limitation to some extent, improvement requires practice. Users must manually control their gaze for input, which may eventually become as simple and seamless as using fingers on a touchscreen. Ultimately, whether another form of input, such as gesture control or voice commands, could enhance usability remains to be seen.

Embedded AR Technologies Beyond Display: play an equally vital role in the success of AR lenses. Compact **power management** and **wireless charging mechanisms** ensure seamless functionality without adding bulk to the lens. For example, Mojo Vision incorporates an ARM processor with a 5GHz radio, medical grade micro-batteries and power management circuitry with wireless recharging components. The lens would communicate directly with a neck-worn device that would act as the companion computer connecting to mobile devices for GPS and wireless connectivity. Motion-tracking sensors, such as accelerometers, gyroscopes, and magnetometers, are critical for ensuring that digital overlays remain stable and aligned with the user's gaze. These sensors enable precise eye tracking, ensuring that augmented visuals move naturally with the wearer's line of sight. Looking ahead, AR lenses are expected to integrate more intuitive control options, such as **gesture recognition** or **voice activation**, reducing reliance on manual gaze control. These advancements aim to provide users with a seamless, hands-free experience, bridging the gap between the physical and digital worlds. By leveraging breakthroughs in sensors, materials, and electronics, AR contact lenses bring together the physical and digital worlds in ways that were previously unimaginable. AR contact lenses represent a convergence of technologies that redefine wearable devices, offering applications beyond vision correction into realms of real-time data interaction and immersive experiences.

7.3 Expanding Healthcare Applications

Contact lenses are emerging as versatile platforms in healthcare, combining vision correction with innovative therapeutic, diagnostic, and optical functionalities. By incorporating advanced materials and technologies, these lenses are transforming the way diseases are monitored, treated, and managed.

7.3.1 Continuous Health Monitoring

Systemic disease plays a critical role in the context of continuous health monitoring, particularly regarding the early diagnosis and management of health conditions. The discovery of robust, selective and specific biomarkers is crucial for the early diagnosis and monitoring of human diseases. The eye is a common target for various conditions, with vision impairment and complications are often associated with both systemic and ocular diseases. Tears, as a bodily fluid in close proximity to the eye, are rich in proteins and other metabolites. They serve as a valuable biomarker repository, offering advantages over other bodily fluids due to their non-invasive collection method [40].

For systemic disease, tear glucose levels provide an important marker for ***diabetes***. The tear film of a patient with diabetes exhibits significantly elevated glucose levels compared to individuals without diabetes, both at fasting and after the peroral glucose load test [41]. Smart contact lenses provide an unprecedented level of continuous tear glucose data acquisition at sub-minute intervals. These advantages allow the precise estimation of lag time, enabling the establishment of the concept called 'personalized lag time'. This approach accounts for individual variations and has been successfully applied to both non-diabetic and diabetic humans, as well as in animal models, demonstrating a high correlation [42]. For instance, smart lenses developed by researchers at Stanford University and South Korea's Pohang University of Science and Technology (POSTECH), can track glucose levels in tears, depicted in Fig. 7.6 [43]. These lenses use glucose presents tears, which undergoes a chemical reaction in the lens's hydrogel material, to generate electricity. Integrated circuits within the lens measure the strength of this current to calculate glucose levels in the tears, and by extension, in the blood. This is a game-changer for diabetes management as users get real-time updates on their blood sugar levels without the need for invasive blood tests.

Tears also hold promise in ***cancer*** diagnostics. The lacrygobin protein, associated with cancer, has been identified in human tears and exhibits high sequence homology to the mammaglobins, which are proteins upregulated in breast cancer and its metastasis. Researchers at the University of New South Wales, Sydney, investigated the utilization of tears for cancer screening [44]. They collected tear samples from patients with various types of cancer and compared to controls. Analysis revealed that lacryglobin was present in the following percentage of patients:

Fig. 7.6 Schematic illustration of the smart contact lens with long term glucose monitoring for contact lens fabricated with the hyaluronated modified gold@platinum (HA-Au@Pt) electrode (adapted with permission from Han et al. [43])

breast cancer (88%), lung (83%), colon (100%), ovary (33%), prostate (100%) and controls (60%). Notably, two control patients with lacryglobin had a family history of breast and prostate cancer. However, lacryglobin was detected in some but not all tear samples, indicating that further studies are needed to explore its potential as a biomarker for cancer.

In ocular health, the SENSIMED Triggerfish® system (Fig. 7.7), which is certified by Conformité Européenne, or European Conformity (CE-marked), and is also FDA approved, provides insights into the ocular volume changes throughout the day and night [45]. The device includes a smart contact lens that captures spontaneous changes in the eye providing physicians with valuable information that can help guide ***glaucoma treatment***. During a typical office visit, a doctor can only assess the eye at a single point in time, which is insufficient for a thorough assessment because the eye changes over a 24-h period in response to pressure, volume, stress and body's normal circadian rhythm (how our body behaves during the day and throughout the night). The SENSIMED Triggerfish® lens is designed to be worn for 24 h, allowing doctors to gather critical information across the entire daily cycle.

7.3.2 Targeted Drug Delivery Systems

Drug-releasing contact lenses represent an exciting advancement in ocular treatment. The first FDA-approved drug-delivering contact lens is Acuvue Theravision with Ketotifen, developed by Johnson & Johnson [46]. These lenses release an antihistamine to treat ocular itch from allergic conjunctivitis, providing fact-acting relief for up to 12 h. These lenses work by releasing the drug into the thin layer of fluid

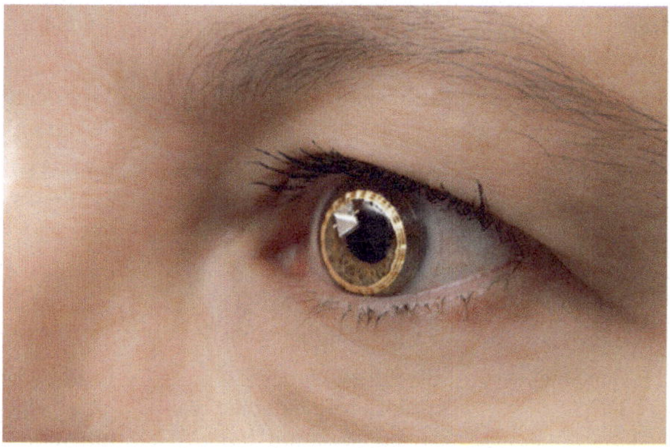

Fig. 7.7 The SENSIMED Triggerfish® Contact Lens (courtesy from SENSIMED)

between the cornea and the contact lens, allows the medication to stay in contact with the eye for an extended period, thereby enhancing its effectiveness. Additionally, the lenses also correct myopia and hyperopia in patients with less than 1.00 D of astigmatism. Beyond treating allergies, drug-delivering contact lenses are being explored for other conditions like glaucoma. Lenses that release medication to lower IOP have shown promise in providing more consistent and effective treatment compared to traditional eye drops.

Dry eye disease (DED), medically referred to as keratoconjunctivitis sicca (KCS) is a condition where drug-releasing contact lenses are showing significant potential. DED is a common disorder of the tear film caused by decreased tear production or increased evaporation, manifesting with a wide variety of signs and symptoms [47]. To address DED, hydrogel-based materials have been developed to detect biomarkers or serve as drug delivery systems, leveraging their excellent biocompatibility, physical and mechanical properties, and long-term implant stability. Interestingly, fluorophores or biosensors can be integrated within hydrogel contact lenses to detect electrolytes in tears for the diagnosis of dry eye [48]. Biosensors made from biocompatible hydrogels can be highly sensitive in diagnosing DED, while these hydrogels also enhance drug bioavailability and retention time, facilitating more effective and long-term treatment. Such so-called smart contact lenses can offer an innovative solution for detecting tear biomarkers or monitoring tear evaporation for a certain period.

The hydrogel-based smart contact lenses in particular, provide clinical potential for detecting DED biomarkers, diagnosing DED, and monitoring its progression. However, integrating batteries into these lenses for long-term power supply remains a challenge due to the limited area of smart contact lenses available. Other limitations include specificity, sensitivity, integration of readout circuits, and repeatability. Once these issues are resolved, biosensors-based system are expected to enable more personalized medical treatment of DED. Additionally, contact lenses may also aid in diagnosing DED by monitoring blink rates, which are often reduced during prolonged near-vision activities such as computer use and reading.

Table 7.1 Examples of drug-releasing contact lenses and their applications in ocular treatments, highlighting recent advancements

Drug/Condition treated	Mechanism	Advancement in drug delivery	References
Timolol for glaucoma	Reduces intraocular pressure by blocking beta-adrenergic receptors in the eye	Embedded within polymer matrices for sustained release over 24 h	[49]
Olopatadine for conjunctivitis	Antihistamine to alleviate ocular itching caused by allergic conjunctivitis	Lenses with advanced molecular imprinting for controlled release	[50]
Dexamethasone for DED	Anti-inflammatory corticosteroid reducing tear film inflammation	Hydrogel-based lenses for improved bioavailability and retention time	[51]
Pirfenidone for Scarring	Antifibrotic drug preventing scarring by inhibiting fibroblast proliferation	Nanostructured carriers integrated into lens material for precise dosing	[52]

Ongoing research focuses on optimizing drug loading capacity and controlled release mechanisms of these lenses as outlined in Table 7.1. These innovations aim to make eye treatments more convenient and effective, potentially improving adherence and outcomes for patients. For example, **timolol-embedded lenses** offer sustained intraocular pressure reduction, improving glaucoma management over traditional eye drops. **Olopatadine-releasing lenses** provide controlled antihistamine delivery, alleviating allergic conjunctivitis symptoms with advanced molecular imprinting techniques. For inflammatory conditions like DED, **dexamethasone-loaded hydrogels** enhance drug bioavailability and retention. Innovations in scarring treatment include **pirfenidone-loaded lenses**, which inhibit fibroblast proliferation through nanostructured carriers, ensuring precise dosing. Similarly, **cyclosporine-delivering lenses** leverage slow-release technology to restore tear film stability. These developments highlight the potential of contact lenses as platforms for targeted, efficient, and patient-friendly drug delivery solutions.

7.3.3 Optics Improvements

Advancements in contact lens technology are revolutionizing optical performance, aiming to address challenges like ***higher-order aberrations*** that affects patient with keratoconus and/or corneal irregularity [53]. New designs and techniques are focused on enhancing visual acuity and providing tailored solutions for these complex optical conditions.

Accommodating contact lenses represent a significant innovation for presbyopia, offering the ability to dynamically transition focus between distant and near vision. These lenses achieve this through advanced mechanisms, such as by applying a small electrical field across liquid crystals or using a mechanical control mechanism [54].

One such design utilizes pressure exerted by the eyelids to squeeze the contact lens and lift its surface [55], resulting in a change in optical power. Another approach redirect fluid within the matrix of the lens to the center, also altering the optical focus [56]. Additionally, accommodating contact lenses have been developed to adjust focus based on the patient's gaze providing seamless and intuitive transitions between near and distant objects [57].

Color vision deficiency (CVD) or color blindness, a congenital ocular disorder, affects approximately one in 12 males and one in 200 females. While no cure exists for CVD, patients opt for wearables like color filtering glasses and lenses to enhance their color perception. A recent innovation by Alam et al. developed 3D-printed contact lenses for CVD management [58]. They synthesized an in situ pHEMA with wavelength-selective filtering dyes. The use of 3D printing offers advantages over conventional techniques, including ease of manufacturing and high dimensional accuracy, making it a suitable platform for the fabrication of customized contact lenses for colorblind individuals.

Color-filtering lenses, such as the X-Chrom (ReVision Optometry) has been helpful for patients with color vision deficiency. The X-Chrom lens is a red-tinted contact lens, placed on one eye to enhance color contrast by darkening yellow-green objects and intensifying red–orange hues [59]. Another innovation, ChromaGen contact lenses features soft lenses available in seven hues and varying densities (light, medium, and dark). According to Alkhudairy and Al Shamlan [60], the ChromaGen lens have been effective in reducing visual stress, improving reading speed and accuracy in patients with dyslexia and Irlen syndrome, even alleviated photosensitivity in patients with Irlen syndrome and cone-rod dystrophy. These lenses also enhanced color vision in patients with color vision deficiency. However, further studies are required to investigate the predictors of improvement and assess the long-term efficacy of ChromaGen lenses on daily activities and learning skills.

These innovations in optics improvements not only enhance everyday visual experiences but also provide practical solutions for individuals with complex refractive errors. By combining advanced materials with adaptive technologies, contact lenses are poised to meet the diverse needs of users, from correcting keratoconus to managing presbyopia dynamically.

7.4 The Future of Vision: Challenges and Opportunities

The previous sections have highlighted remarkable examples of what the future holds with the advancement of contact lens technology, from disease detection, management and treatment of various ocular diseases. It is no surprise that the constant evolution of circuit miniaturization, alongside innovation in nanotechnology and 3D printing has sped up progress in smart contact lens research. However, these advancements come with their own set of challenges.

The **Technical challenges** in smart contact lens development extend beyond miniaturization and integration to include achieving key intrinsic properties. Lenses must maintain **optical power, biocompatibility, softness, transparency, wettability, and oxygen transmissibility** to ensure comfort and clear vision for users. Another crucial requirement is **overnight wearability**, allowing continuous IOP monitoring even during sleep. Advances in material science are needed to support **ergonomic curvature fitting** for different corneal shapes and sizes, ensuring high measurement accuracy and within-subject repeatability. Additionally, lenses must withstand **mechanical and chemical durability** to endure daily cleaning, disinfecting, and handling for routine, long-term use. These features are essential for user acceptance and comfort, but incorporating them into sophisticated, electronics-embedded designs poses a significant challenge. User acceptance relies heavily on **ergonomic curvature fitting** that accommodates diverse corneal shapes and sizes for accurate measurements and repeatability. Additionally, ensuring **overnight wearability** for applications like continuous IOP monitoring during sleep is crucial. Another key challenge is**data security and privacy**, as these lenses collect sensitive health information. Ensuring **secure data transmission** through robust encryption protocols will be critical to maintaining user trust. Lastly, **wireless communication** between lenses and external devices, such as smartphones or medical monitors, must be reliable and interference-free. This requires advancements in **antenna design and signal transmission technologies** to achieve stable, seamless connectivity.

The **Safety, Regulatory, and Compliance Challenges** further complicate the path to commercialization. **Stringent testing and approval** processes are required to ensure the safety and efficacy of these lenses, given their intimate contact with the eye. Evaluating **biocompatibility and long-term effects** is essential to mitigate risks and protect users' ocular health. Additionally, ethical considerations surrounding **data privacy**, **informed consent**, and the potential misuse of technology must be addressed. Ensuring **health and safety** involves preventing infections, maintaining oxygen permeability, and avoiding adverse reactions to materials. **Regulatory approval** from bodies such as the FDA requires rigorous testing and validation to meet safety and efficacy standards. Privacy concerns also remain prominent; securing data collected by the lenses, including biometric information, is vital for user trust. Lastly, **environmental sustainability** is becoming increasingly important. While companies are exploring biodegradable and environmentally friendly materials for disposable lenses, performance challenges with these materials remain a significant hurdle.

The **Future Innovations and Impact** of smart lenses hold immense promise. AR and virtual reality (VR) capabilities could revolutionize gaming, entertainment, and professional training. Additionally, advanced lenses may play a role in **prosthetics and vision restoration**, offering solutions for individuals with certain types of blindness. In the long term, contact lenses could potentially act as non-invasive **brain-computer interfaces**, enabling seamless interaction between humans and external devices. Despite the challenges, the continuous advancements in this field pave the way for smart lenses to transform healthcare, user experience, and our interaction with the world.

The journey of contact lens technology, as explored in this book, reflects a remarkable evolution from simple vision correction tools to sophisticated, multifunctional devices poised to revolutionize healthcare and daily life. Through advancements in material science, nanotechnology, and electronics, contact lenses now offer possibilities far beyond traditional applications, including real-time health monitoring, drug delivery, AR, and vision restoration. However, this progress comes with challenges such as ensuring safety, affordability, environmental sustainability, and regulatory compliance. Collaborative efforts between researchers, industry leaders, and healthcare providers are crucial to overcoming these hurdles and driving widespread adoption. As the boundaries of innovation continue to expand, contact lenses are set to redefine wearable technology, offering transformative solutions that bridge the gap between healthcare and the digital age, enhancing both quality of vision and quality of life.

References

1. Leonardi, M., Leuenberger, P., Bertrand, D., Bertsch, A., Renaud, P.: A soft contact lens with a MEMS strain gage embedded for intraocular pressure monitoring. In: TRANSDUCERS '03. 12th International Conference on Solid-State Sensors, Actuators and Microsystems. Digest of Technical Papers (Cat. No.03TH8664), vol. 2, pp. 1043–1046 (2003)
2. Rabensteiner, D.F., Rabensteiner, J., Faschinger, C.: The influence of electromagnetic radiation on the measurement behaviour of the triggerfish® contact lens sensor. BMC Ophthalmol. **18**, 338 (2018). https://doi.org/10.1186/s12886-018-1013-x
3. Lingley, A.R., Ali, M., Liao, Y., Mirjalili, R., Klonner, M., Sopanen, M., Suihkonen, S., Shen, T., Otis, B.P., Lipsanen, H.: A single-pixel wireless contact lens display. J. Micromechanics Microengineering. **21**, 125014 (2011). https://doi.org/10.1088/0960-1317/21/12/125014
4. Yao, H., Shum, A.J., Cowan, M., Lähdesmäki, I., Parviz, B.A.: A contact lens with embedded sensor for monitoring tear glucose level. Biosens. Bioelectron. **26**, 3290–3296 (2011). https://doi.org/10.1016/j.bios.2010.12.042
5. Labiotech: What Happened to the Plans for a Smart Contact Lens for Diabetics? (2018)
6. Sifferlin, A.: Google Granted Patent for Smart Contact Lens, https://time.com/3758763/google-smart-contact-lens/
7. Kaur, S., Kim, Y.-J., Milton, H., Mistry, D., Syed, I.M., Bailey, J., Novoselov, K.S., Jones, J.C., Morgan, P.B., Clamp, J.: Graphene electrodes for adaptive liquid crystal contact lenses. Opt. Express **24**, 8782–8787 (2016). https://doi.org/10.1364/OE.24.008782
8. Park, J., Ahn, D.B., Kim, J., Cha, E., Bae, B.-S., Lee, S.-Y., Park, J.-U.: Printing of wirelessly rechargeable solid-state supercapacitors for soft, smart contact lenses with continuous operations. Sci. Adv. **5**, eaay0764 (2019). https://doi.org/10.1126/sciadv.aay0764
9. Deshpande, A., Ghosh, C., Karkhanis, M.U., Banerjee, A., Pourshaban, E., Hasan, M.R., Nikeghbal, A., Dastgir, M.G., Kim, H., Mastrangelo, C.H.: Automatic light intensity modulation using TNC-based artificial iris for smart contact lens. In: 2023 IEEE Sensors, pp. 1–4. IEEE (2023). https://doi.org/10.1109/SENSORS56945.2023.10325096
10. Fortune Business Insights: Contact Lenses Market Size, Share & Industry Analysis, By Modality (Reusable and Disposable), By Design (Toric, Multifocal, and Spherical), By Material (Soft, Gas Permeable {Orthokeratology and Scleral}, and Others), By Distribution Channel (Ophthalmologists, Retail Stores, and Online Stores), and Regional Forecast, 2024–2032 (2024)
11. Fleming, R.: The Evolution of Smart Contact Lenses, https://westtechfest.com/the-evolution-of-smart-contact-lenses/
12. Moreddu, R., Vigolo, D., Yetisen, A.K.: Contact Lens Technology: From Fundamentals to Applications (2019). https://doi.org/10.1002/adhm.201900368

13. Xia, Y., Khamis, M., Fernandez, F.A., Heidari, H., Butt, H., Ahmed, Z., Wilkinson, T., Ghannam, R.: State-of-the-art in smart contact lenses for human–machine interaction. IEEE Trans. Human-Machine Syst. **53**, 187–200 (2022). https://doi.org/10.1109/THMS.2022.3224683
14. Zhu, H., Yang, H., Zhan, L., Chen, Y., Wang, J., Xu, F.: Hydrogel-based smart contact lens for highly sensitive wireless intraocular pressure monitoring. ACS Sens. **7**, 3014–3022 (2022). https://doi.org/10.1021/acssensors.2c01299
15. Yang, C., Wu, Q., Liu, J., Mo, J., Li, X., Yang, C., Liu, Z., Yang, J., Jiang, L., Chen, W., Chen, H.J.: Intelligent wireless theranostic contact lens for electrical sensing and regulation of intraocular pressure. Nat. comm. **13**, 2556 (2022). https://doi.org/10.1038/s41467-022-29860-x
16. Bobrowski, T., Arribas, E.G., Ludwig, R., Toscano, M.D., Shleev, S., Schuhmann, W.: Rechargeable, flexible and mediator-free biosupercapacitor based on transparent ITO nanoparticle modified electrodes acting in μM glucose containing buffers. Biosens. Bioelectron. **101**, 84–89 (2018). https://doi.org/10.1016/j.bios.2017.10.016
17. Kim, J., Cha, E., Park, J.: Recent advances in smart contact lenses. Adv. Mater. Technol. **5**, 1900728 (2020). https://doi.org/10.1002/admt.201900728
18. Kim, T.Y., Shin, S., Choi, H., Jeong, S.H., Myung, D., Hahn, S.K.: Smart contact lenses with a transparent silver nanowire strain sensor for continuous intraocular pressure monitoring. ACS Appl. Bio Mater. **4**, 4532–4541 (2021). https://doi.org/10.1021/acsabm.1c00267
19. Leonardi, M., Leuenberger, P., Bertrand, D., Bertsch, A., Renaud, P.: First steps toward noninvasive intraocular pressure monitoring with a sensing contact lens. Invest. Ophthalmol. Vis. Sci. **45**, 3113–3117 (2004). https://doi.org/10.1167/iovs.04-0015
20. Ren, H., Zheng, L., Wang, G., Gao, X., Tan, Z., Shan, J., Cui, L., Li, K., Jian, M., Zhu, L.: Transfer-medium-free nanofiber-reinforced graphene film and applications in wearable transparent pressure sensors. ACS Nano **13**, 5541–5548 (2019)
21. Xu, J., Cui, T., Hirtz, T., Qiao, Y., Li, X., Zhong, F., Han, X., Yang, Y., Zhang, S., Ren, T.-L.: Highly transparent and sensitive graphene sensors for continuous and non-invasive intraocular pressure monitoring. ACS Appl. Mater. Interfaces **12**, 18375–18384 (2020). https://doi.org/10.1021/acsami.0c02991
22. Lee, H., Kim, S., Kim, K.-B., Choi, J.-W.: Scalable fabrication of flexible thin-film batteries for smart lens applications. Nano Energy **53**, 225–231 (2018). https://doi.org/10.1016/j.nanoen.2018.08.054
23. Lin, B., Wang, M., Zhao, C., Wang, S., Chen, K., Li, X., Long, Z., Zhao, C., Song, X., Yan, S.: Flexible organic integrated electronics for self-powered multiplexed ocular monitoring. NPJ Flex. Electron. **6**, 77 (2022). https://doi.org/10.1038/s41528-022-00211-6
24. Yun, J., Li, Z., Miao, X., Li, X., Lee, J.Y., Zhao, W., Lee, S.W.: A tear-based battery charged by biofuel for smart contact lenses. Nano Energy **110**, 108344 (2023). https://doi.org/10.1016/j.nanoen.2023.108344
25. Chiou, J.-C., Hsu, S.-H., Huang, Y.-C., Yeh, G.-T., Liou, W.-T., Kuei, C.-K.: A Wirelessly Powered Smart Contact Lens with Reconfigurable Wide Range and Tunable Sensitivity Sensor Readout Circuitry (2017). https://doi.org/10.3390/s17010108
26. Kim, J., Kim, M., Lee, M.-S., Kim, K., Ji, S., Kim, Y.-T., Park, J., Na, K., Bae, K.-H., Kyun Kim, H.: Wearable smart sensor systems integrated on soft contact lenses for wireless ocular diagnostics. Nat. Commun. **8**, 14997 (2017). https://doi.org/10.1038/ncomms14997
27. Musgrave, C.S.A., Fang, F.: Contact lens materials: a materials science perspective. Materials (Basel) **12**, 261 (2019). https://doi.org/10.3390/ma12020261
28. Sweeney, D.F.: Clinical signs of hypoxia with high-Dk soft lens extended wear: is the cornea convinced? Eye Contact Lens **29**, S22–S25 (2003). https://doi.org/10.1097/00140068-200301001-00007
29. Shaker, L.M., Al-Amiery, A.A., Kadhum, A.A.H., Takriff, M.S.: Manufacture of contact lens of nanoparticle-doped polymer complemented with zemax. Nanomaterials **10**, 2028 (2020)
30. Lee, M.-J., Park, S.-Y., Sung, A.-Y.: Characterization of biocompatible hydrogel lenses using methacrylic acid with neodymium oxide nanoparticles. Polymers (Basel) **13**, 1575 (2021)

31. Shaker, L.M., Abdulhadi, S., Al-Azzawi, W.K., Alamiery, A., Takriff, M.S., Isahak, W.N.R.W.: Colorless poly (vinyl pyrrolidone) hydrogel contact lenses synergized with silver nanoparticles. J. Opt. **53**, 847–856 (2024). https://doi.org/10.1007/s12596-023-01176-z
32. Shaker, L.M., Al-Azzawi, W.K., Al-Amiery, A., Takriff, M.S., Wan Isahak, W.N.R.: Highly transparent antibacterial hydrogel-polymeric contact lenses doped with silver nanoparticles. J. Vinyl Addit. Technol. **29**, 1023–1035 (2023). https://doi.org/10.1002/vnl.21995
33. Shaker, L.M., Alamiery, A., Takriff, M., Wan Isahak, W.N.R.: Novel blue-wavelength-blocking contact lens with Er_3+/TiO_2 NPs: manufacture and characterization. Nanomaterials **11**, 2190 (2021). https://doi.org/10.3390/nano11092190
34. Sirviö, J.A., Visanko, M.: Highly transparent nanocomposites based on poly (vinyl alcohol) and sulfated UV-absorbing wood nanofibers. Biomacromol **20**, 2413–2420 (2019). https://doi.org/10.1021/acs.biomac.9b00427
35. Zhao, W., Zhu, Y., Zhang, J., Xu, T., Li, Q., Guo, H., Zhang, J., Lin, C., Zhang, L.: A comprehensive study and comparison of four types of zwitterionic hydrogels. J. Mater. Sci. **53**, 13813–13825 (2018). https://doi.org/10.1007/s10853-018-2535-6
36. Alió, J.L., Plaza-Puche, A.B., Piñero, D.P., Amparo, F., Rodríguez-Prats, J.L., Ayala, M.J.: Quality of life evaluation after implantation of 2 multifocal intraocular lens models and a monofocal model. J Cataract Refract Surg **37**, 638–648 (2011). https://doi.org/10.1016/j.jcrs.2010.10.056
37. Bohac, M., Jagic, M., Biscevic, A., Lukacevic, S., Mravicic, I., Suic, S.P., Dekaris, I.: Stereoacuity and multifocal intraocular lenses—a systematic review. Acta Inf. Med. **31**, 62–67 (2023). https://doi.org/10.5455/aim.2023.31.62-67
38. Shan, Y., Hua, J., Mao, H.: 3D printing of optical lenses assisted by precision spin coating. Adv. Funct. Mater. **34**, 2407165 (2024). https://doi.org/10.1002/adfm.202407165
39. Bastian, M.: AR contact lens: this is the new Mojo Lens prototype, https://mixed-news.com/en/ar-contact-lens-mojo-vision-unveils-new-prototype/
40. Nandi, S.K., Singh, D., Upadhay, J., Gupta, N., Dhiman, N., Mittal, S.K., Mahindroo, N.: Identification of tear-based protein and non-protein biomarkers: its application in diagnosis of human diseases using biosensors. Int. J. Biol. Macromol. **193**, 838–846 (2021). https://doi.org/10.1016/j.ijbiomac.2021.10.198
41. Sen, D.K., Sarin, G.S.: Tear glucose levels in normal people and in diabetic patients. Br. J. Ophthalmol. **64**, 693–695 (1980). https://doi.org/10.1136/bjo.64.9.693
42. Park, W., Seo, H., Kim, J., Hong, Y.-M., Song, H., Joo, B.J., Kim, S., Kim, E., Yae, C.-G., Kim, J., Jin, J., Kim, J., Lee, Y., Kim, J., Kim, H.K., Park, J.-U.: In-depth correlation analysis between tear glucose and blood glucose using a wireless smart contact lens. Nat. Commun. **15**, 2828 (2024). https://doi.org/10.1038/s41467-024-47123-9
43. Han, H.H., Kim, S.-K., Kim, S.-J., Choi, I., Mok, J.W., Joo, C.-K., Shin, S., Hahn, S.K.: Long-term stable wireless smart contact lens for robust digital diabetes diagnosis. Biomaterials. **302**, 122315 (2023). https://doi.org/10.1016/j.biomaterials.2023.122315
44. Evans, V., Vockler, C., Friedlander, M., Walsh, B., Willcox, M.D.P.: Lacryglobin in human tears, a potential marker for cancer. Clin. Experiment. Ophthalmol. **29**, 161–163 (2001). https://doi.org/10.1046/j.1442-9071.2001.00408.x
45. FDA: FDA permits marketing of device that senses optimal time to check patient's eye pressure, https://www.fda.gov/news-events/press-announcements/fda-permits-marketing-device-senses-optimal-time-check-patients-eye-pressure
46. Johnson & Johnson: Johnson & Johnson Vision Care Receives FDA Approval for ACUVUE® Theravision^{TM} with Ketotifen—World's First and Only Drug-Eluting Contact Lens, https://www.jjvision.com/press-release/johnson-johnson-vision-care-receives-fda-approval-acuvuer-theravisiontm-ketotifen
47. Phadatare, S.P., Momin, M., Nighojkar, P., Askarkar, S., Singh, K.K.: A comprehensive review on dry eye disease: diagnosis, medical management, recent developments, and future challenges. Adv. Pharm. **2015**, 704946 (2015). https://doi.org/10.1155/2015/704946
48. Li, Q., Cao, Y., Wang, P.: Recent Advances in Hydrogels for the Diagnosis and Treatment of Dry Eye Disease (2022). https://doi.org/10.3390/gels8120816

49. Yang, H., Zhang, F., Fan, Y., Zhang, J., Fang, T., Xing, D., Zhen, Y., Nie, Z., Liu, Y., Wang, D., Li, J.: Co-delivery of brinzolamide and timolol from micelles-laden contact lenses: in vitro and in vivo evaluation. Pharm. Res. **41**, 531–546 (2024). https://doi.org/10.1007/s11095-024-03672-8
50. Xue, Y., Zhang, W., Lei, Y., Dang, M.: Novel polyvinyl pyrrolidone–loaded olopatadine HCl–laden doughnut contact lens to treat allergic conjunctivitis. J. Pharm. Sci. **109**, 1714–1724 (2020). https://doi.org/10.1016/j.xphs.2020.01.022
51. Bengani, L.C., Kobashi, H., Ross, A.E., Zhai, H., Salvador-Culla, B., Tulsan, R., Kolovou, P.E., Mittal, S.K., Chauhan, S.K., Kohane, D.S., Ciolino, J.B.: Steroid-eluting contact lenses for corneal and intraocular inflammation. Acta Biomater. **116**, 149–161 (2020). https://doi.org/10.1016/j.actbio.2020.08.013
52. Panthagani, J., Vincent, R., Butt, G.F., Wallace, G.R., Rauz, S.: The effect of Pirfenidone on TGF induced profibrotic pathways—a potential for treating corneal fibrosis. Invest. Ophthalmol. Vis. Sci. **61**, 2608 (2020)
53. Jones, L., Hui, A., Phan, C.-M., Read, M.L., Azar, D., Buch, J., Ciolino, J.B., Naroo, S.A., Pall, B., Romond, K.: BCLA CLEAR–Contact lens technologies of the future. Contact Lens Anterior Eye **44**, 398–430 (2021). https://doi.org/10.1016/j.clae.2021.02.007
54. Bailey, J., Morgan, P.B., Gleeson, H.F., Jones, J.C.: Switchable liquid crystal contact lenses for the correction of presbyopia. Crystals **8**, 29 (2018). https://doi.org/10.3390/cryst8010029
55. Win-Hall, D.M.: Accommodating or focusable contact lens (2009)
56. Waite, S., Gupta, A., Payor, R., Christ, R., Christ, M.D.: Sacrificial molding process for an accommodating contact lens (2017)
57. Biederman, W.J., Yeager, D.J., Otis, B., Pletcher, N.: Capacitive gaze tracking for auto-accommodation in a contact lens (2016)
58. Alam, F., Salih, A.E., Elsherif, M., Yetisen, A.K., Butt, H.: 3D printed contact lenses for the management of color blindness. Addit. Manuf. **49**, 102464 (2022). https://doi.org/10.1016/j.addma.2021.102464
59. Siegil, I.M.: The X-Chrom lens. On Seeing Red. Surv. Ophthalmol. **25**, 312–324 (1981). https://doi.org/10.1016/S0039-6257(81)80001-X
60. Alkhudairy, Z., Al Shamlan, F.: The use of chromagen lenses in different ocular and non-ocular conditions: a prospective cohort study. Cureus. **14**, (2022). https://doi.org/10.7759/cureus.28963

Index

© The Editor(s) (if applicable) and The Author(s), under exclusive license to Springer Nature Singapore Pte Ltd. 2025
N. Adrus et al., *Contact Lenses*, Series in BioEngineering,
https://doi.org/10.1007/978-981-96-3636-5

The manufacturer's authorised representative in the EU is Springer Nature Customer Service Centre GmbH, Europaplatz 3, 69115 Heidelberg, Germany. If you have any concerns regarding our products, please contact ProductSafety@springernature.com

Printed and bound by CPI Group (UK) Ltd, Croydon, CR0 4YY
01/10/2026
02225936-0003